A Capabilities-Based Strategy for Army Security Cooperation

Jennifer D. P. Moroney, Adam Grissom, Jefferson P. Marquis

Prepared for the United States Army

ARROYO CENTER

The research described in this report was sponsored by the United States Army under Contract No. DASW01-01-C-0003.

Library of Congress Cataloging-in-Publication Data

Moroney, Jennifer D. P., 1973–
    A capabilities-based strategy for Army security cooperation / Jennifer D.P.
Moroney, Adam Grissom, Jefferson P. Marquis.
        p. cm.
    Includes bibliographical references.
    ISBN 978-0-8330-4199-9 (pbk. : alk. paper)
    1. Combined operations (Military science) 2. United States. Army. 3. Military
planning—United States—Methodology. 4. Multinational armed forces—
Organization. 5. Military assistance, American. 6. United States—Military
relations—Foreign countries. 7. Military art and science—International cooperation.
I. Grissom, Adam. II. Marquis, Jefferson P. III. Title.

U260.M67 2007
355.4'6—dc22

                                                                                                    2007030770

The RAND Corporation is a nonprofit research organization providing objective analysis and effective solutions that address the challenges facing the public and private sectors around the world. RAND's publications do not necessarily reflect the opinions of its research clients and sponsors.

RAND® is a registered trademark.

Cover photo by Michael A. Osur.

© Copyright 2007 RAND Corporation

Published 2007 by the RAND Corporation
1776 Main Street, P.O. Box 2138, Santa Monica, CA 90407-2138
1200 South Hayes Street, Arlington, VA 22202-5050
4570 Fifth Avenue, Suite 600, Pittsburgh, PA 15213-2665
RAND URL: http://www.rand.org/
To order RAND documents or to obtain additional information, contact
Distribution Services: Telephone: (310) 451-7002;
Fax: (310) 451-6915; Email: order@rand.org

Preface

This report documents research conducted for the U.S. Army on the feasibility of adopting a "niche capabilities" approach to multinational force compatibility (MFC) with non-allied armies. It is the latest in a series of RAND Arroyo Center studies, from a project entitled "A Capabilities Based Strategy for Army Security Cooperation," supporting the Army's efforts to bolster MFC with the spectrum of potential coalition partners.

It is increasingly difficult for the Army to ensure compatibility with the range of armies with which it may be asked to operate in the future. Army Transformation is introducing new capabilities that cannot be duplicated by most potential coalition partners. At the same time, unpredictable *ad hoc* "coalitions of the willing" are replacing alliances as the dominant form of multinational operations. Many members of these coalitions are less capable than traditional allies, and they often have little history of cooperation with the U.S. Army. Preparing to operate with them is therefore challenging.

Army planners working on the problem of multinational force compatibility require a planning framework to guide and focus the service's long-term compatibility investments with these less capable armies. This report defines and describes such a framework, which we call the Niche Capability Planning Framework. It provides a conceptual template for integrating the various considerations, ranging from U.S. Army capability gaps, to the politics of collaborating with foreign armies, to the coordination of Army and Department of Defense security cooperation activities, implicit in a strategy for cultivating compatible niche capabilities in non-core partner armies.

The research for this report was sponsored by the Deputy Chief of Staff, G-35 (Strategy, Plans and Policy), and was conducted in RAND Arroyo Center's Strategy, Doctrine, and Resources Program. RAND Arroyo Center, part of the RAND Corporation, is a federally funded research and development center sponsored by the United States Army.

The Project Unique Identification Code (PUIC) for the project that produced this document is DAPRRY015.

For more information on RAND Arroyo Center, contact the Director of Operations (telephone 310-393-0411, extension 6419; FAX 310-451-6952; email Marcy_Agmon@rand.org), or visit Arroyo's web site at http://www.rand.org/ard/.

Contents

Figures

Tables

Summary

This study outlines a planning framework for cultivating multinational force compatibility (MFC) with armies that are not traditional allies. Such coalition partners are increasingly important to the Army in the post-9/11 security environment. Multilateral military operations are often now conducted by "coalitions of the willing" rather than by alliances, and many of these *ad hoc* coalitions include key contingents that have no history of sustained peacetime cooperation with the U.S. Army. The Army has only very limited resources available to enhance compatibility with non-allied partner armies, especially compared to the resources devoted to compatibility with traditional allies such as the United Kingdom. The challenge of enhancing compatibility and building partnership capacity with non-core partner armies therefore requires an innovative approach to planning.

Niche Capabilities: A New Approach to MFC Planning for Army Security Cooperation

This study outlines an approach to MFC planning that focuses Army resources on potential coalition partners that possess, or with assistance could possess, niche capabilities that would augment U.S. Army capabilities in useful ways. The approach, which we term the Niche Capability Planning Framework, features a four-phase planning cycle. The first phase focuses on characterizing and prioritizing candidate niche capabilities to meet potential Army capability shortfalls. With global responsibilities and limited resources, the Army is obliged to accept risk in its investments in future capabilities. This risk is manifested in decisions to eschew investment in capabilities that the Army desires but cannot afford and, also, in decisions to acquire limited quantities of some capabilities with the risk that there may not be enough of these capabilities under certain circumstances.[1] Some of these potential gaps and shortfalls are likely to be more important than others, due to the nature of the capability and the context in which it may be used. Some may also be easier than others to develop in partner armies. Capabilities in which the Army has decided not to invest, but which are nevertheless potentially important to the Army's effectiveness in a future contingency and cost-effective to develop in non-core partner armies, would be ideal niche capabilities. The first phase of the Niche Capability Planning Framework focuses on identifying such niche shortfalls or gaps.

The second phase of the Niche Capability Planning Framework focuses on assessing potential partner armies to identify those suited to becoming niche contributors. In seeking potential non-core partners from among the nearly 190 armies worldwide, Army planners must gauge the political acceptability, willingness to cooperate, and existing capabilities of each. Armies that are eager to work with the United States will make better niche partners than those that are ambivalent. Armies that are in good standing with the U.S. government, from the perspective of the broader bilateral political relationship, will make better niche partners than those with which the U.S. government maintains a contentious relationship. Armies that already possess a desirable niche capability, or the means to support a capability cultivated with U.S. assistance, will be better niche partners than those that do not. Phase two of the Niche Capability Planning Framework is devoted to analyzing these issues.

The third phase of the Niche Capability Planning Framework focuses on the development of an integrated program of military-to-

---

[1] The recent cancellations of the Comanche scout helicopter and the Crusader artillery system are examples of the first type of decision. The Army's traditional reluctance to support a large active component Civil Affairs structure is an example of the second type of decision.

military contacts, security assistance, and other security cooperation activities designed to cultivate capabilities in selected partner armies. Ideally, this process will leverage Army security cooperation resources with the Department of Defense (DoD) and Interagency security cooperation resources to accomplish the Army's objectives at the least cost to the service and the nation.

The fourth and final phase of the Niche Capability Planning Framework is the execution of the phased program, coupled with continuing assessment and the development of appropriate measures of effectiveness. This will allow the Army Staff to ensure that its plans are being implemented and that lessons are learned and incorporated into the plan over time. We envision the Niche Capability Planning Framework as a cycle that will be repeated periodically by Army planners.

Recommendations

We recommend that the Army incorporate a strategy for working with non-core partner armies into the Army Security Cooperation Strategy (ASCS). We further recommend that Army planners adopt a deliberate planning framework for designing and implementing a niche capabilities strategy. This framework should, at a minimum, incorporate analyses of projected Army capability gaps, key partner characteristics, and the resources required to match capabilities with partners in practice. This report outlines such an approach in the form of the Niche Capability Planning Framework.

Acknowledgments

The authors owe a great debt to a number of officers, civil servants, and analysts for their assistance on this study. These include current and past members of Army Staff G-35, U.S. European Command, U.S. Army Europe, V Corps Headquarters, Special Operations Command Europe, U.S. Central Command, U.S. Army Central Command, U.S. Southern Command, U.S. Army South, U.S. Pacific Command, U.S. Army Pacific, Special Operations Command Pacific, the Joint Staff, the Office of the Secretary of Defense, and the National Guard Bureau.

At RAND, Tom Szayna, Lauri Zeman, and Olga Oliker provided valuable comments on early drafts. Jasen Castillo contributed significantly to the intellectual development of the framework.

The authors are also most grateful for the thoughtful and insightful feedback from our reviewers, Dr. Douglas Lovelace from the Army War College and Dr. Kevin McCarthy from RAND.

The project officer for this study was Mr. Mark McDonough, Chief of the Multinational Force Compatibility Directorate in Army Staff G-35. Mr. McDonough provided outstanding support to the study on both substantive and administrative matters. We are grateful for his guidance and help throughout this two-year effort.

Abbreviations

ABCA	American-British-Canadian-Australian Armies Standardization Program
AOR	Area of Responsibility
AR	Army Regulation
ARCENT	U.S. Army Central Command
ARTEP	Army Training and Evaluation Program
ASA	Assistant Secretary of the Army
ASA/ALT	Assistant Secretary of the Army (Acquisition, Logistics, and Technology)
ASCC	Army Service Component Command
ASCS	Army Security Cooperation Strategy
CATS	Combined Arms Training System
CENTCOM	U.S. Central Command
CMEP	Civil-Military Emergency Planning
COCOM	Combatant Command
CONUS	Continental United States
DoD	Department of Defense

DOTMLPF	Doctrine, Organization, Training, Materiel, Leadership, Personnel, and Facilities
DTRA	Defense Threat Reduction Agency
EDA	Excess Defense Articles
EUCOM	European Command
FM	Field Manual
FMF	Foreign Military Financing
GTEP	Georgia Train and Equip Program
HD/LD	High-Demand/Low-Density
HMMWV	High-Mobility Multi-Purpose Wheeled Vehicle
HQDA	Headquarters Department of the Army
IMET	International Military Education and Training
IPL	Integrated Priority List
JCTP	Joint Contact Team Program
MACOM	Major Command
METL	Mission Essential Task List
MFC	Multinational Force Compatibility
MP	Military Police
MRX	Mission Rehearsal Exercise
MTOE	Modified Table of Organization and Equipment
MTP	Mission Training Plan
MTT	Mobile Training Teams
OIF	Operation Iraqi Freedom
OSD	Office of the Secretary of Defense

PCS	Permanent Change of Station
PfP	Partnership for Peace
POM	Program Objective Memorandum
QDR	Quadrennial Defense Review
SCG	OSD Security Cooperation Guidance
SPG	Strategic Planning Guidance
SPP	State Partnership Program
TOE	Table of Organization and Equipment
TRADOC	U.S. Army Training and Doctrine Command
TSC	Theater Security Cooperation
TSCMIS	Theater Security Cooperation Management Information System
TSCP	Theater Security Cooperation Plan
USAREUR	U.S. Army Europe
USARPAC	U.S. Army Pacific
USARSO	U.S. Army South
USG	U.S. Government

CHAPTER ONE

Introduction

The U.S. Army is confronted by major challenges as it seeks to enhance multinational force compatibility (MFC) with potential coalition partners.[1] During the Cold War, the Army focused primarily on conventional warfighting operations in familiar locales.[2] Key partner armies were linked tightly to the U.S. Army by longstanding relationships.[3] The Army possessed a clear idea of where potential operations would probably occur, which multinational partners would most likely participate, and what type of military operations the coalition or alliance would conduct. This yielded a relatively stable MFC planning environment.

In the post-9/11 era, the Army faces a much less predictable MFC planning environment. The Army cannot confidently forecast the location or nature of future contingencies, which may range from benign peace operations in the western hemisphere to theater-level conventional warfighting on the other side of the globe. Furthermore, planners cannot know which armies, beyond a handful of consistent allies,

---

[1] The Army defines MFC as "the collection of capabilities, relationships, and processes that together enable the Army to conduct effective multinational operations across the full spectrum of military missions. It encompasses not only the capability to conduct effective military operations with coalition partners, but also the factors that contribute to the development and maintenance of an alliance or coalition relationship," Army Regulation 34-1, *Army Multinational Force Compatibility*, 2004, p. 2-1.

[2] Primarily Western Europe, Korea, and Southwest Asia.

[3] Primarily NATO, Major Non-NATO Allies (MNNAs) Australia, South Korea, and Japan, and a number of partners in Southwest Asia.

might participate in future coalitions. The MFC planning challenge is therefore becoming more complex.

The increased complexity and unpredictability of the international security environment, combined with the fact that the technical, tactical, and operational sophistication of U.S. forces is far outpacing that of most potential coalition partners, require that Army planners develop an Army-wide MFC planning framework to focus and rationalize the resources available for building compatibility with partner armies. The framework would optimally promulgate Army MFC priorities throughout the service and the broader community of Department of Defense (DoD) and U.S. government (USG) agencies conducting security cooperation with foreign governments.

The purpose of this study is to assist the Army in developing a planning framework for cultivation of MFC with less capable partners. This category of armies, which we term "non-core partners"[4] to distinguish them from traditional or highly capable allied armies, is increasingly important in the context of the shifting "coalitions of the willing" of the post-9/11 era.[5] However, the resources available to enhance compatibility with this category of armies are quite limited, especially compared to resources devoted to MFC with traditional allies. The challenge of enhancing compatibility with non-core partner armies therefore requires an innovative approach to planning.

Study Objectives and Tasks

This study has two objectives. The first is to develop a planning framework for cultivating MFC with non-core partner armies. The second

---

[4]   We define non-core partner armies as prospective coalition partners that are neither formal allies nor close regional partners with quasi-ally status. The primary indicator of "non-core" status is the lack of a stable long-term program of collaborative MFC-focused Army security cooperation activities aimed at producing tactical- and operational-level compatibility.

[5]   The term "coalition of the willing" refers to an *ad hoc* political-military grouping that exists outside formal international institutions and is focused on accomplishing an objective that is limited in time and scope. The phrase has often been associated with the George W. Bush Administration, but it originated in the early 1990s.

is to help the Army gain greater visibility into MFC-oriented Army security cooperation activities, DoD security cooperation programs, combatant command (COCOM) theater security cooperation planning methodologies, and USG security cooperation programs that are relevant to the Army's approach to MFC planning.

Organization of the Study

This monograph comprises seven chapters and an appendix. Chapter Two describes the strategic rationale for cultivating niche capabilities in non-core partner armies. It begins with a discussion of the challenges facing MFC planners and the policy framework within which they operate. It then outlines the current process for Army MFC planning. It concludes with a description of the proposed Niche Capability Planning Framework.

Chapter Three describes an approach to identifying candidate niche capabilities. It incorporates four assessment criteria: complementarities with U.S. Army needs, potential operational contributions, practicability, and dual-utility for the partner nation.

Chapter Four describes an approach to identifying promising niche partners. It begins with a discussion of the problem of collective action outside the confines of a formal alliance relationship. It then outlines the framework for selecting candidate countries for niche capability cultivation. The framework provides a filter for categorizing candidate partner armies according to three criteria: political acceptability, availability for coalition operations, and military capability.

Chapter Five describes an approach to identifying the training and equipment required to develop a niche capability. It parallels the Army's own Combined Arms Training System for readying units, building from mission statements to essential tasks to training requirements.

Chapter Six translates the approach detailed in Chapter Five into the security cooperation sphere, where Army security cooperation activities will be used to cultivate niche capabilities in non-core partner armies. It seeks to help Army planners maximize each activity and to leverage the programs and resources available from DoD and other USG agencies. It begins with an overview of key security cooperation concepts, such as funding source, initiative, program, and activity, that are often confused among planners at Headquarters, Department of the Army (HQDA), the Office of the Secretary of Defense (OSD), COCOMs, and in the field. It then presents a systematic approach to phasing these resources to cultivate niche capabilities in non-core partner armies.

Chapter Seven consolidates and presents the team's recommendations. It suggests that the Army consider adopting the Niche Capability Planning Framework as a component of the overall Army Security Cooperation Strategy. It also suggests a number of planning, policy, and programmatic changes that would, in our view, make Army MFC planning more effective.

The appendix defines and describes key concepts in the security cooperation arena, many of which are often misunderstood by planners and policymakers, to provide a clear conceptual basis for the Niche Capability Planning Framework.

CHAPTER TWO

Conceptualizing Multinational Force Compatibility

The U.S. Army will be facing many significant challenges as it contemplates future coalition operations. Recent experience provides evidence that armies with greatly varying capabilities are, and will likely continue to be, vital members of coalitions conducting operations with the United States. It also underscores the difficulty of operating alongside non-core partner armies, particularly if there is little history of prior cooperation.

Cultivating multinational force compatibility is one of the Army's primary responsibilities. Title 10 of the U.S. Code instructs the Army to organize, train, and equip its units to meet the warfighting requirements of joint force commanders. One key enduring warfighting requirement, as stated in recent Defense Planning Guidance, is the ability to operate with multinational partners. The Army must therefore prepare itself and, in many cases, prepare partner armies to operate together on future battlefields. This function is fulfilled through Army security cooperation activities.

This chapter reviews the rationale for cultivating niche capabilities with non-core partner armies. We first discuss the Army MFC planning process and situate MFC in the broader context of DoD security cooperation planning. We then detail the proposed Niche Capability Planning Framework.

The MFC Planning Context

The U.S. Army has traditionally focused the majority of its MFC resources on a relatively fixed set of allies, such as the United Kingdom, Canada, Australia, and Germany. This approach has merit. These allies maintain highly capable armies, traditionally help shoulder the burden in U.S.-, NATO-, and UN-led operations, and are likely to participate in many future operations. However, the U.S. Army is increasingly required to provide forces to joint commanders conducting multinational operations with a more diverse group of coalition partners, ranging from highly capable allies to marginally capable partner armies with little history of cooperation with the U.S. Army.[1] There is thus a need to pursue focused, high-payoff security cooperation with a broader range of potential coalition partners to build compatibility and improve the prospects for success in future operations.

In the current planning context, a primary challenge for the Army will be to identify the right partner armies and Army security cooperation activity combinations to accomplish this objective while working within resource constraints. Our research suggests that a capabilities-based approach to planning MFC with non-core partners would help planners manage this challenge. Before describing the new framework, however, we first review the current framework for MFC planning.

Army MFC planning occurs in a complex political-military milieu. MFC-oriented activities are a subset of Army security cooperation activities, which are in turn a subset of DoD security cooperation activities. They are simultaneously operational activities and diplomatic activities, important both as political symbols and as "phase zero" of future war plans and contingency operations. As such, they have a high profile, and the MFC planning process is subject to many layers of policy and planning guidance.

---

[1] The coalitions conducting operations in Afghanistan, Iraq, and the Balkans are examples.

Army-Level Security Cooperation Guidance and the ASCS

The National Security Strategy (NSS), Defense Strategy (DS), National Military Strategy (NMS), Strategic Planning Guidance (SPG), Quadrennial Defense Review (QDR), and the OSD Security Cooperation Guidance (SCG) set the stage for the conduct of DoD security cooperation activities.

The Army operationalizes the security cooperation planning guidance it receives from OSD (SCG) and the COCOM Theater Security Cooperation Plans (TSCPs) in three documents: the Army Security Cooperation Strategy (ASCS), the Multinational Force Compatibility (MFC) Plan, and Army Regulation (AR) 34-1, *Multinational Force Compatibility*.[2]

Army planners use the MFC-related guidance in the ASCS to prioritize efforts to build MFC with partner armies. To date, this linkage has been implicit. In 2004 the Army Staff decided to issue a formal MFC Plan to make this linkage more explicit and transparent.[3]

The MFC Plan

The purpose of the MFC Plan, when published, will be "to provide implementation guidance to the Army in the identification, prioritization, integration, and assessment of those activities that enhance its ability to operate with and effectively integrate contributions from non-U.S. forces and non-traditional actors across a full range of military missions."[4] Currently in draft form, the MFC Plan has the potential to help planners understand their role in the broader DoD security cooperation and ASCS frameworks. It also establishes MFC goals that are derived from the ASCS.

---

[2] Army Regulation AR 34-1 does not change on an annual basis like the OSD SCG and the COCOM TSCP do.

[3] Following the completion of the research for this study, the G-35 decided to shift the focus on the MFC Plan to interoperability. The Army is presently evaluating the separate publication of this plan.

[4] G-35/SSI, "MFC Guidance in Progress Review (IPR)," 31 March 2004.

The Army participates in MFC activities to build effective coalitions for the full range of military operations. This implies that the Army has two overarching MFC objectives: *promoting interoperability* between the United States and select allies and partners, and *building capabilities* that will enhance the effectiveness of future coalition operations. In a perfect world, the U.S. Army would be fully compatible with all of the other 190-plus armies of the world, across any conceivable type of operation, in any region, against any type of adversary, with little or no advance notice. Constraints on time and resources make that impossible. The MFC Plan and indeed the ASCS serve to help the Army prioritize its available resources.

As it develops, the MFC Plan will likely provide planning guidance for building compatibility with the entire range of partner armies, from longstanding allies to members of "coalitions of the willing." It will rationalize MFC planning across Army components to ensure that the service receives maximum benefit from its MFC investment. This study seeks to help the Army Staff as it develops its MFC Plan, particularly as it addresses MFC building with non-core partner armies.

MFC Administration: AR 34-1

The third key MFC planning document is Army Regulation 34-1, *Multinational Force Compatibility.* It defines and describes the Army's process for planning, implementing, and monitoring MFC-oriented activities. The Army Staff revised AR 34-1 in 2003 to better position the service to work effectively with coalition partners in the new global security environment. Several important changes resulted. Most noticeably, the title of AR 34-1 was changed from *Military Rationalization, Standardization, and Interoperability* to *Multinational Force Compatibility,* a seminal shift in terminology.[5] The new regulation defines multinational force compatibility as

---

[5] The term MFC contrasts with the old concept of rationalization, standardization, and interoperability (RSI). Where RSI focused mostly on materiel-technical issues, MFC-oriented activities focus on the overarching goal of achieving compatibility with key allies and partners in a broad set of operational categories, and are supposed to represent a holistic approach to improving the performance of multinational coalition operations.

the collection of capabilities, relationships, and processes that together enable the Army to conduct effective multinational operations across the full spectrum of military missions. It encompasses not only the capability to conduct effective military operations with coalition partners, but also the factors that contribute to the development and maintenance of an alliance or coalition relationship.[6]

Guidance provided in AR 34-1 applies to the active Army, the Army National Guard, and the Army Reserve. According to AR 34-1,

- The scope and focus of MFC-building activities and supporting activities are tailored to the specific military mission of the alliance or coalition and to the roles of the participating nations' governmental and/or nongovernmental organizations, international organizations, and/or informal non-state military organizations;
- In addition to enhancing the operational effectiveness (through improved interoperability) of an alliance or coalition, Army security cooperation activities may contribute to the creation and maintenance of alliances or coalitions; and
- MFC achievement must be measurable and will be subject to analytical assessment on a regular basis.

AR 34-1 also describes the process used to plan the Army's efforts to build MFC. As described above, this process is complex and involves a number of key players within and outside the Army.

- **HQDA responsibilities.** The G-35/SSI has oversight authority but does not directly control all of the specific activities conducted to build MFC. The office is responsible for a variety of tasks, to include promulgating MFC policy and priorities such as eliminating duplication/redundancies among MFC-oriented Army security cooperation activities, and integrating and disseminating

---

[6] AR 34-1, para. 2-1.

regional combatant commander and institutional Army priorities for MFC to the responsible Army commands and agencies.[7]

- **ASA/ALT responsibilities.** The Assistant Secretary of the Army (Acquisition, Logistics, and Technology) (ASA/ALT) also has an integral role to play in the development of MFC with key allies and partners. For example, ASA/ALT is responsible for incorporating MFC considerations and requirements in Army-wide technology base strategy, policy, guidance, planning, and acquisition programs.[8]

- **Other MFC players.** In addition to ASA/ALT, AR 34-1 specifies responsibilities for a number of other actors involved in MFC planning.[9] Each of these offices, agencies, and commands plays an important role—often as a subject-matter expert—in MFC activities and is therefore involved in MFC planning.

Army MFC planning also involves, directly and indirectly, a number of actors outside the Army. The regional COCOMs, the Office of the Secretary of Defense, various DoD agencies, the State Department, and a variety of other USG offices and authorities issue requirements for Army security cooperation activities or provide resources and conduct activities that support or complement Army programs. To maximize the achievements of these activities and resources, it is crucial for these actors to coordinate closely their planning and implementation. To date, this coordination process is largely informal, a situation addressed in detail in Chapter Six.

MFC-Oriented Army Security Cooperation Activities

The Army pursues MFC through a range of Army security cooperation activities, most of which have been traditionally directed at key allies,

---

[7] AR 34-1, p. 7.

[8] AR 34-1, p. 9.

[9] These actors include the Army Chief Information Officer, the Army Deputy Chief of Staff for Intelligence (G-2), the Army Deputy Chief of Staff for Logistics (G-4), the Army Chief of Engineers, the Judge Advocate General, U.S. Army Training and Doctrine Command, U.S. Army Materiel Command, U.S. Army Forces Command, regional component commands, and heads of delegation/ABCA national points of contact.

such as the ABCA (American-British-Canadian-Australian Armies Standardization Program) countries. The following are examples of the Army security cooperation activities that build MFC:

1. **NATO forums.** The U.S. Army participates in several NATO forums whose primary purpose is the enhancement of MFC. These include the Military Committee, NATO Committee for Standardization, NATO Standardization Agency, Military Committee Land Standardization Board, Conference of National Armaments Directors, NATO Army Armaments Group, Joint C3 Requirements and Concepts Subcommittee, Senior NATO Logisticians' Conference, and Land Electronic Warfare Working Group.

2. **ABCA Armies' Program.** A quadripartite agreement between the American, British, Canadian, and Australian armies that ensures interoperability, the highest degree of cooperation between the armies through materiel and nonmateriel standardization. ABCA features a number of working groups on key MFC topics and produces handbooks that codify the state of the art and are used by a large number of the world's armies to plan coalition operations.

3. **Five Power forums.** Focused interoperability discussions including the armies of the United Kingdom, France, Germany, Italy, and the United States in the Senior National Representative (Army) (SNRA) process.

4. **Army-to-Army staff talks and bilateral meetings.** Bilateral and multilateral discussions that occur between G-35 and army staff leaders in foreign countries. In such meetings, goals and priorities for Army security cooperation activities for the coming year are discussed.[10]

---

[10] In developing an overall strategy for staff talks, specific goals include: ensuring that staff talks address current and future strategic, operational, and tactical security challenges that will confront both the U.S. Army and likely allies and coalition partners; establishing priorities to ensure that staff talks enhance mutual understanding and influence the development of future battlefield requirements; contributing to the ability to conduct combined operations; integrating the results of staff talks into pertinent U.S. Army programs in sup-

In addition, according to the AR 34-1, there are numerous "MFC-related" activities that, while not directly conducted or administered by HQDA, serve to enhance the Army's ability to operate effectively in a multinational coalition. These include, but are not limited to, the following types of activities:[11]

- Educational exchanges, to include foreign students at U.S. Army schools and U.S. Army students at foreign schools;
- Personnel and unit exchanges, liaison officers, and visits of senior military and civilian officials;
- Regional combatant commander and Army component commander sponsored forums;
- Combined exercises and operations;
- Regional Army programs;
- International cooperative research, development, and acquisition (RDA);
- Purchase of material designed and/or produced by other countries;
- Transfer of materiel designated and/or produced by the United States to its allies and coalition partners;
- Multinational logistics;
- Foreign Area Officer program; and
- Cooperative religious, moral, morale, and ethical support activities conducted with chaplain corps of allies and coalition partners.

In sum, the Army MFC planning framework is complex. Policy guidance flows from the national level to the Department of Defense and then to the Army, where there is broader guidance for Army security cooperation planning and more specific guidance for building MFC. A variety of actors are involved in MFC planning, both inside and outside the Army, and a variety of security cooperation activities can be brought to bear on the MFC challenge. To date, this complex

---

port of Army transformation and the global war on terrorism; and establishing management metrics.

[11] See AR 34-1 for detailed descriptions of each MFC and MFC-related activity.

planning system has tended to treat non-core partner armies in an *ad hoc* manner. The next section describes a niche capabilities approach to MFC planning that has the potential to be a major improvement over the current approach.

Proposed Niche Capability Planning Framework

This section outlines an approach to focusing Army MFC resources on non-core partners that possess, or could with assistance possess, niche capabilities that would augment U.S. Army capabilities in useful ways. This framework, which we term the Niche Capability Planning Framework, features a four-phase planning cycle.

The first phase of the planning cycle focuses on characterizing and prioritizing candidate niche capabilities to meet potential Army capability shortfalls. The U.S. Army is an extraordinarily capable organization, perhaps the most powerful of its type in existence, but it does nevertheless have relative deficiencies and shortfalls. Two types of shortfalls are particularly interesting from the MFC planning perspective. The first type comprises capabilities that the Army has decided not to create in its own force structure, for political or budgetary reasons. We call these shortfalls "capability niches." The second type of niche occurs where the Army has decided to accept some risk in its inventory of capabilities by acquiring a capability in an amount that will not be sufficient to meet the operational requirement under all circumstances. As a result, the Army sometimes finds itself facing a shortage of these capabilities. Some refer to these as "high-demand/low-density" (HD/LD) capabilities; we call them "capacity niches" because the Army lacks capacity in this portion of its inventory.

The second phase of the new planning framework is to assess non-core armies to identify appropriate partners. This assessment centers on three questions. First, is the potential partner politically acceptable to the United States?[12] Second, is the potential partner favorably disposed

---

[12] Political acceptability is partially determined by the strategic significance the United States places on the candidate state. For example, a more strategically significant state may

to working with the United States on international security challenges? Moreover, is it in favor of developing a niche capability, particularly the one the United States has in mind for it? Third, does the potential partner possess the overall defense structure required to support a niche capability cultivated with the United States?

The third phase of the Niche Capability Planning Framework is the development of a phased program of military-to-military contacts, security assistance, combined exercises, and other Army security cooperation activities to cultivate the identified capabilities in the selected partner armies. This phase requires HQDA to work closely with OSD and the COCOMs to determine security cooperation requirements and priorities. This process will allow for the identification of niche capabilities the Army will seek to cultivate and the partners with which it proposes to cooperate.

The fourth phase features close coordination between HQDA and major subordinate commands, DoD elements, and USG agencies to focus Army, DoD, and Interagency resources on these priority capabilities and partners. The Army faces a shortage of resources for building MFC, so success will require strong focus and coordination among the disparate players in the security cooperation arena.[13]

We believe that the Niche Capability Planning Framework could improve Army planning in at least four important ways. First, the Army cannot possibly build capabilities with all 190-plus armies of the world, nor even with the entire force structures of a limited number of non-core partners. In many cases, the current approach spreads Army security cooperation resources thinly across non-core partner armies (both across large numbers of partner armies and across large segments of each partner army's force structure), thereby failing to create real capability that is useful during operations. The niche capabilities approach recognizes that some partners are more useful coalition members than others and, perhaps more importantly, that some non-core partner

---

be found to be acceptable, even though it lacks the mature and institutionalized democratic processes of a less strategically significant state.

[13] Memoranda of Agreement/Understanding might be a useful tool to facilitate coordination among the various stakeholders.

capabilities are more useful to the Army than others, due primarily to the relative strengths and weaknesses of the U.S. Army's own force structure. The Niche Capability Planning Framework would enable the Army to focus its scarce MFC resources on non-core partner capabilities that best complement its existing suite of capabilities—either relieving HD/LD stresses or adding unique capabilities that are not in the Army's existing repertoire of capabilities.

Second, the niche capabilities approach to building multinational force compatibility would allow the Army, in coordination with the Army service component commands (ASCCs) and COCOMs, to conduct discussions with potential non-core partners regarding their likely participation in future coalition operations. In some cases, this advance groundwork would make it easier for Army, DoD, and USG planners to predict which countries will participate in particular operations and, just as important, the capabilities that non-core partners are likely to offer when they do choose to participate in an operation. The niche capabilities strategy may thereby help planners manage the unpredictability of the contemporary security environment.

Third, the niche capabilities strategy will also help the Army meet COCOM and DoD security cooperation requirements in a manner that maximizes benefits in terms of Army Title 10 requirements. Much of the Army's current slate of activities is driven by COCOM and OSD political-military objectives. This is appropriate. However, in some cases the Army contribution to COCOM and OSD security cooperation is conducted in a manner that does not help the Army with its responsibility to improve MFC. For example, many Army security cooperation activities focus on capabilities that do not complement Army capabilities or are not sequenced to create real compatibility between the partner and the U.S. Army. In a few cases, this may be necessary due to political-military considerations. Most of the time, however, it is simply due to a lack of awareness among Army, OSD, and COCOM planners of the potential compatibility payoff that might occur if activities were focused and sequenced appropriately. Often, with foresight, it should be possible to meet OSD and COCOM political-military objectives while simultaneously creating compatibilities that are useful to the Army. The niche capabilities strategy provides a framework for identifying and exploiting such synergies.

Finally, the niche capabilities strategy will help the Army think about compatibility with non-core partners in a proactive and systematic manner. The niche approach is not a panacea. To the extent the Army must depend on niche capabilities possessed by potential coalition partners, its freedom of action can be circumscribed by this dependence. The more indispensable the potential partner's niche capability, the more vulnerable the United States is to manipulation by adversaries and, potentially, partners as well. Therefore, the most unique and important niche capabilities should be reserved for the most reliable coalition partners.

Moreover, even limiting MFC cooperation to a few specific niche capabilities, the Army cannot work with every possible non-core partner. Nor will every potential partner want to work with the Army in peacetime. As a result, even if the Army implements the niche approach aggressively, some multinational contingents that show up for future coalition operations will be neither longstanding allies nor niche capability partners. This will continue to present an enduring challenge to the Army, a challenge that previous RAND research suggests will have to be managed largely through internal Army efforts to make its tactical forces more flexible in coalition contexts.[14] The Niche Capability Planning Framework can also contribute by making Army planners aware of the relative shortfalls in the Army's inventory of capabilities, and the types of capabilities maintained by non-core partner armies. Together, these may allow Army warfighters and war planners to identify potential coalition contributions.

While there are grounds to believe that a niche capabilities strategy for MFC is both promising and feasible, there are several analytical and policy challenges the Army will need to consider. These include, among others, determining precisely what capabilities can or should be augmented by non-core partner armies (in close coordination with the

---

[14] Adam Grissom, Nora Bensahel, John Gordon, Terrence K. Kelly, and Michael Spirtas, *U.S. Army Transformation and the Future of Coalition Warfighting*, Santa Monica, CA: RAND Corporation, 2006.

COCOMs and ASCCs), providing effective guidance to the ASCCs and other major commands (MACOMs) that execute Army activities, deconflicting within the Interagency to avoid duplication and identify gaps, improving interoperability of the niche across the DOTMLPF, ensuring that the sequencing of Army security cooperation activities is optimized to increase absorption rates, and maintaining the capabilities over time. These factors are discussed in greater detail in Chapter Six and in the concluding chapter of this report.

CHAPTER THREE

Identifying Candidate Niche Capabilities

The Niche Capability Planning Framework takes as its starting point the needs of the U.S. Army. The purpose of this chapter is to outline a conceptual framework for identifying candidate niche capabilities that would, if cultivated successfully in non-core partner armies, enable the U.S. Army to fulfill its assigned role in the Defense Strategy with less risk.

Approach

As discussed in Chapter Two, the Army's planning environment is uncertain and its MFC resources are scarce. If it did not face resource constraints, the U.S. Army would be able to invest unlimited effort and resources into building capabilities in each of the world's armies. In this ideal world, the Army would be fully prepared to operate in any portion of the conflict spectrum with whatever partners might be produced by shifting coalitions. Clearly, resource realities do not allow this approach, and never will.

Instead, Army MFC planners require a framework to focus the service's efforts on particularly promising opportunities for cooperation. Proven allies possessing advanced full-spectrum capabilities clearly top the list. Beyond this tiny group, however, the planning challenge becomes much more complex. The Army currently lacks a conceptual approach to identifying promising opportunities for cooperation with non-core partner armies. This chapter describes such an approach.

The Niche Capability Planning Framework is rooted in Army Title 10 responsibilities. The framework seeks to identify areas in which the Army has chosen to accept risk in its investments in capabilities for future operations. Risk is a central element of Army force planning, broadly defined as the likelihood that the Army will be unable to accomplish its assigned objectives in some future operation or that it will be obliged to accomplish its objectives at a cost (measured in terms of lives and resources) greater than it might have been. With global responsibilities and limited resources, the Army is constantly forced to balance risk across its investments in future capabilities.

To strike this balance, Army planners attempt to predict future requirements. These requirements are a function of the types of operations the service will conduct in the future, the types of adversaries they will be conducted against, and the types of Army capabilities required to conduct them successfully. There is considerable uncertainty in each of these areas. As a result, the Army spreads its investments across those combinations it believes are most likely and most important to U.S. interests. Where these investments fall short of potential requirements, the Army accepts a degree of operational risk.

Broadly speaking, there are two approaches to identifying areas of operational risk. One approach applies inductive reasoning to the recent past. Planners gather lessons-learned data from recent and ongoing operations or measure the tempo of Army units to identify which elements of the force have been insufficient in the recent past. From these past shortfalls they inductively derive future shortfalls. The strength of this approach is that the results have empirical validity and are rooted in actual experience. The weakness of the inductive approach is that it assumes the future will resemble the past. Should the future not mirror the Army's past experience, if the Army begins conducting new types of operations or is confronted by new types of adversaries, then the inductive approach may lead planners to misidentify requirements and, thereby, potential gaps in capabilities.

As a result, the Army also utilizes deductive assessments of future requirements. Planners posit a set of premises and assumptions about future operations, analyze the match between these requirements and existing Army capabilities, and deductively derive potential shortfalls.

Formal war games and constructive modeling are the most common types of deductive analysis. Prominent examples include the Total Army Analysis, the Capability Gaps analysis performed by Training and Doctrine Command (TRADOC), the DoD-wide Operational Availability series of analyses, the Integrated Priority Lists submitted by the combatant commands and their Army components, and the wide-ranging constructive modeling conducted by the TRADOC Analysis Center in support of the acquisition and programming offices of the Army Staff.

There are two primary advantages to deductive analysis. First, the premises of the analysis can be modified to reflect predicted changes in the nature of future operations. Second, the potential impact of a given capability can be measured with some precision through multiple excursions of the analysis. For example, a capability can be added to a constructive model of a future operation to determine whether it affects the operation's outcome. After running the model with the new capability, planners can discern whether the capability improves the effectiveness of friendly forces.

However, forward-looking analyses also have important weaknesses. Most prominent among these is the possibility that the premises and assumptions at the heart of the model, war game, or analysis may not be correct. If future operations have characteristics different from those posited by the analysts, then the Army may have misidentified its requirements and areas of operational risk.

Ideally, Army planners using the Niche Capability Planning Framework would employ both inductive and deductive analysis to identify potential capability gaps. They would also leverage, to the greatest extent possible, existing analyses in order to minimize the burden of implementing the planning framework. Shortfalls identified by both inductive analysis and deductive analysis are more likely to prove to be useful niche capabilities than those highlighted by one type of analysis but not another.

Once planners have developed a sense of the Army's gaps, these gaps must be prioritized. Logically, this prioritization must take two considerations into account. First, how important is each gap? This can be analyzed using both inductive and deductive analyses, as described above. Second, how likely is it that MFC-oriented activities can contribute to closing a gap? Both factors must be considered together in order to focus MFC planning in areas that are both important and MFC-appropriate.[1]

After planners have identified gaps and prioritized them according to operational importance and MFC-appropriateness, they will want to consider a final factor: the partners. MFC cooperation is a two-way street, involving the interests of both partners. Focusing MFC-oriented activities on Army gaps makes little sense if partner armies will not agree to participate. As a result, in many cases it will be appropriate to give priority to focus areas that will be attractive to potential partner armies (and governments).

In sum, a number of factors must be taken into account when developing MFC-oriented Army security cooperation activities. To capture these considerations in a useful format, we have developed a formal planning framework for selecting candidate niche capabilities. This framework considers four criteria: complementarity, operational contribution, practicability, and dual-utility.

Criterion 1: Complementarity

The first and most important criterion is the degree to which a candidate niche capability matches a real or potential shortfall in the Army's own capabilities. Put another way: Is there a niche in the U.S. Army into which the candidate coalition partner capability might fit? Our research suggests that two types of niches may exist.

First, there are "capability niches." A coalition partner might provide a capability that is entirely absent from the U.S. Army inventory of capabilities. The U.S. Army is full spectrum but not comprehensive, and while it maintains a breadth of capabilities well in excess of any other contemporary army, it cannot maintain a full suite of every capa-

---

[1]   Of course, both evaluations are subjective. Operational importance could be assessed through wargaming or analyses of lessons learned, but ultimately it will require a subjective military judgment of what capabilities are likely to be required in future operations, and how important each type of capability will be to success. Assessing MFC-appropriateness is also a subjective judgment, but one more amenable to the expertise of Army planners and implementers.

bility that might prove useful in a contingency. Paramilitary constabulary or "carabinieri" forces and humanitarian demining capabilities are two examples of capabilities that the Army has chosen not to develop. These kinds of capabilities provide potential coalition partners with an opportunity to augment a gap in U.S. Army capabilities. Moreover, there may be some capabilities the Army may wish to forgo, or divest itself of, because they can be acquired more cost-effectively and, perhaps, in a more politically acceptable manner through the niche capability approach.

Second, there are "capacity niches." Despite the efforts of service planners to forecast the types of capabilities required in future operations, some capabilities have much more demand than can be accommodated by the current force structure. These high-demand/low-density capabilities have presented a serious challenge to the Army throughout the post–Cold War era. Operations in Afghanistan and Iraq have made the problem more severe, requiring "more people in more places for more time."[2] Recently, the Army has moved to address the worst of the HD/LD mismatch by proposing to reallocate 100,000 soldiers to HD/LD specialties, but there is a widespread recognition that the HD/LD problem will persist.[3] The expansive nature of U.S. national interests, reflected in the "1-4-2-1" Defense Strategy (protect the homeland, deter forward in four critical regions, swiftly defeat adversaries in two near-simultaneous regional conflicts, decisively defeating one of them) virtually ensures that the Army will be forced to accept some risk in certain types of capabilities. These potential shortfalls present opportunities for potential coalition partners to contribute to the depth of U.S. land power. Civil affairs and special operations are potential examples of "capacity niches" within the Army's inventory of capabilities.

---

[2] Bruce Nardulli, "The U.S. Army and the Offensive War on Terrorism," in Lynn Davis and Jeremy Shapiro (eds.), *The U.S. Army and the New National Security Strategy*, Santa Monica, CA: RAND Corporation, 2003, p. 33.

[3] For details on the 100,000 reallocation, see internal Army figures quoted in Michael O'Hanlon, "Rebuilding Iraq and Rebuilding the U.S. Army," Saban Center Middle East Memo Number 3, 4 June 2004. Last accessed 27 October 2004 at:
http://www.brookings.edu/views/op-ed/ohanlon/20040604.htm

Criterion 2: Operational Contribution

The second criterion for assessing potential niche capabilities is the potential operational significance of the capability. This criterion captures the capability's importance to the overall effectiveness of the coalition and to the likelihood of campaign success. It is possible to imagine a virtually infinite set of niche capabilities that meet the first criterion, complementarity, without necessarily adding much to the overall effectiveness of the coalition. An assessment of operational contribution, while necessarily rough and subject to debate, is therefore crucial to the niche capabilities assessment framework.

Operational contribution can be stated in positive or negative terms: as enhancement of a coalition's operational effectiveness or as mitigation of operational risk.[4] Traditionally, the emphasis was always placed on conventional combat operations and the classic effectiveness criteria of lethality, survivability, and mobility. But it is clear, examining the evidence from the major operations of the past 15 years, that the U.S. Army has achieved a remarkable overmatch in conventional combat operations. This experience has also demonstrated that stability operations are equally vital to overall campaign success, and that many operations, regardless of size and initial conditions, will involve stability and perhaps reconstruction operations at some stage. The relative value and, therefore, the emphasis placed on conventional combat operations and stability and reconstruction operations will vary according to assessment of the strategic and operational environments. In making such assessments, Army planners might consider the scenarios set forth in the Secretary of Defense's Strategic Planning Guidance. An assessment of the potential operational contribution of a niche capability might, therefore, grant equal importance to combat and stability and reconstruction operations.

Without the resources of detailed wargaming and field experimentation, analysts are left to estimate the operational contribution of a niche capability via three methods. First, analysts can look to formal service-collected lessons learned from recent operations. Second, they

---

[4]   U.S. Department of Defense, *Annual Defense Review*, Washington, D.C.: Government Printing Office, 2004. See especially Chapter 5.

can wargame future scenarios to project the contribution of alternative capabilities. Third, they can use counterfactual analysis to impute the potential contribution a niche capability might have made in a past operation. Each of these methods is subjective, and none will produce certain answers. Nevertheless, they help to structure debate over alternative niche candidates and provide at least a heuristic indicator of potential operational contribution.

Criterion 3: Practicability

Pragmatically speaking, some niche capabilities are easier than others to cultivate and sustain. Equally, some coalition capabilities are easier than others to integrate into a multinational force. Both factors must be incorporated into any overall assessment of candidate niche capabilities.

When assessing ease of cultivation and sustainment, conceptual complexity and technological intensity are important indicators. Niche capabilities that are highly complex may require large procurement expenditures, highly trained officers and enlisted soldiers, and high levels of operations and maintenance expenditure. Missile defense and precision strike might, for example, be poor candidate niche capabilities for these reasons. Some non-core partner armies may be able to sustain such capabilities, but most will not. Additionally, the very limited resources provided for Army security cooperation activities that promote MFC are unlikely to succeed in cultivating complex and expensive capabilities. On balance, then, less complex and technologically intensive niche capabilities would be preferable.

When assessing ease of coalition integration, the most important consideration may be the nature of the candidate niche capability itself. Some capabilities are relatively modular and easily understood by coalition partners, and therefore less difficult to integrate effectively into a multinational force. Others are highly unusual, and therefore more difficult to integrate. Additionally, some capabilities, if they are to be used effectively, must be tightly integrated with the rest of the coalition. Air defense is perhaps an example of this kind of capability. Other capabilities require less integration at the tactical and operational levels, or at least the ramifications of loose coordination are less dire. The coalition infantry units occupying some of the quieter sectors of Iraq and Afghanistan are examples. On balance, candidate niche capabilities that are easy to integrate into a multinational force would be preferable to those that are not.

Moreover, it is important to note that niche capabilities provided by most non-core partner armies will require substantial, if not total, combat support and combat service support from American forces during an operation. The relative need for such support is another consideration in evaluating the practicability of a candidate capability.

Criterion 4: Dual-Utility

The final criterion to be considered is the utility of the candidate niche capability in the domestic context of the partner country. Some non-core partners are highly developed and stable societies, but the majority of non-core partners are developing states with greater or lesser degrees of political stability.

Some candidate niche capabilities, if cultivated successfully in non-core partner armies, could conceivably make an important contribution to stability in the partner country. Others, especially high-tech capabilities for conventional warfighting, are much less likely to make such a contribution.[5]

On balance, those candidate capabilities that would make such a contribution would be preferable to those that would not. Two factors lie behind this judgment. First, from a purely functional perspective, the capabilities that contribute to the stability of the partner country are correspondingly more likely to be available during a contingency. However, this cuts both ways, and the greater the dual-utility of a niche capability to a candidate partner, the less likelihood of its being available for external use in a contingency situation.

Second, niche capabilities that contribute to state stability will better serve the overall aims of U.S. foreign policy than those that do not. They correspondingly deserve to receive priority. Along the same

---

[5] Though there are potential exceptions, such as sophisticated theater-wide communications systems that might be used for both military and civil purposes.

lines, capabilities that might easily lead to destabilization or repression in the partner country would not be good candidates for cultivation.

Applying the Approach

In applying this approach for selecting candidate niche capabilities, planners would use each of the four criteria as an independent filter or gate. Resource constraints will oblige the Army to focus on a very few capabilities, perhaps three to five at any given time. The primary purpose of the four criteria is therefore to "weed out" all capabilities that would be inappropriate for any reason. Those capabilities that survive this selection process should be strong in all four areas. Figure 3.1 presents this process in a visual form. In the appendix the criteria are applied to the contemporary security environment to generate a short list of potential niche capability candidates.

The application of the capability selection framework would logically proceed in a phased manner. The analysis should begin by focusing on the Army's own Title 10 requirements, namely the complementarity criterion. This will ensure that no effort is expended analyzing capabilities that are of no use to the Army. Army security cooperation planners are fortunate that at present there are a number of ongoing

Figure 3.1
Candidate Selection Framework

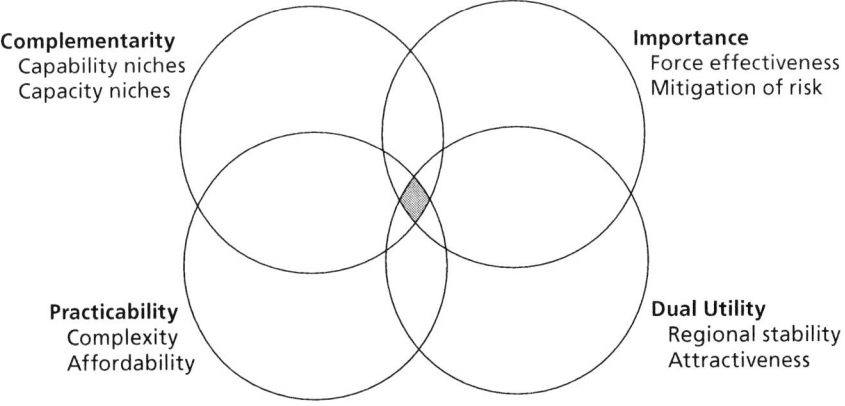

RAND *MG563-3.1*

analytical efforts within the service that they can leverage. Where capability niches are concerned, the TRADOC Army Capabilities Integration Center (ARCIC) is leading an Army-wide effort to identify, characterize, and prioritize capability gaps in the current and programmed force.[6] As a minimum standard, Army planners might stipulate that a candidate niche capability must be reflected on TRADOC's list of capability gaps.[7]

In terms of capacity niches, at the time of this study, DoD was conducting the Operational Availability 2006 (OA-06) in support of the 2006 Quadrennial Defense Review. The Operational Availability analyses are the department's primary tool for determining whether the services' existing structures are sufficient to cover all the operations they might plausibly be required to conduct. In particular, OA-06 incorporates a Baseline Security Posture that defines a steady-state profile of contingencies over the next several years. Joint planning teams have developed force templates for each of the contingencies. These requirements are aggregated over time to measure the sufficiency of service structures at any given point over the multiyear period. Army planners can use these results to identify capacity niches, perhaps stipulating at a minimum that the capability be stressed according to the OA-06 analysis.[8]

The second criterion considered should be the operational importance of the capability, to identify those that may make a significant contribution to the effectiveness of the Army in the field. Here again the capability gaps analysis being conducted by ARCIC can be of service to Army planners. TRADOC has prioritized capability gaps according to a rigorous process. This might be used by Army planners to measure the operational contribution of a candidate niche capability. Alternatively, Army planners can consider the Strategic Planning Guidance scenarios and the Integrated Priority Lists (IPLs) issued yearly by the

---

[6] This analysis is sensitive in nature, limiting what can be said in this venue.

[7] Another possible source of insight is the Joint Capabilities Integration and Development System (JCIDS) managed by the Joint Staff, though this tends to be focused on materiel rather than broader capabilities.

[8] The OA-06 analysis is classified.

combatant commands, which offer a prioritized set of capabilities these commanders believe are underresourced. Finally, Army planners might consider utilizing constructive modeling to measure the operational importance of a candidate niche capability. At a minimum, Army planners might stipulate that a candidate niche capability must appear on the TRADOC list of capability gaps, a COCOM IPL, or be shown in constructive modeling to measurably enhance the operational success of friendly forces.

The third step would consider the practicability of cultivating the niche capability, which is partly a measure of raw feasibility and partly a measure of U.S. resources required to cultivate the capability in a non-core partner. Army planners should analyze each candidate capability to identify the minimum training and equipment required to cultivate it. Chapter Five provides a framework for doing so. At a minimum, Army planners might stipulate that it should take no longer than six months to train partner forces for the niche capability and that the equipment should require no advanced individual functional training to maintain (i.e., electronics courses). The appendix illustrates how the Army might conduct an assessment of cost-effectiveness.

The final step should be to evaluate the capabilities on the basis of dual-utility, in order to identify those most likely to be attractive to potential non-core partners. Army planners can use history as a guide for this step, assessing whether a candidate capability has been used in the past to counter internal threats to stability. One or more historical examples should be sufficient to demonstrate the potential dual-utility of the candidate capability.

Applying the framework in this phased manner will ensure that planners consider and apply the Army's priorities first, before considering larger questions of feasibility and attractiveness to potential partners. Our assessment is that the most promising niche capabilities will tend to be combat support capabilities relevant to the later phases of stability operations. The U.S. Army has traditionally emphasized investments in combat arms capabilities for conventional conflict. As a result, capability and capacity shortfalls both tend to occur in stability operations capabilities. Within this set, combat support and combat service support capabilities tend to be relegated to the National Guard and Army Reserve, which by policy can rotate less often, meaning that capacity niches tend to be found in these capabilities. Among this set, combat support capabilities tend to be more immediately and obviously vital to operational success than combat service support capabilities, meaning that combat support tends to rank higher in this metric. Finally, the key dual-use capabilities for internal security tend to be combat support capabilities such as intelligence, military police and constabulary, civil affairs, and training teams. As a result, according to our assessment, promising niche capabilities will tend to cluster in combat support functions applicable to stability operations.

Conclusion

Army planners will be challenged by an uncertain international security environment in their attempts to identify, select, and defend candidate niche capabilities for further investment under the ASCS. The approach proposed in this chapter is intended to assist in this process. The assessment must begin within the Army, by identifying which types of capabilities would be most useful as it conducts operations under today's extraordinary conditions. Practicability and dual-utility considerations are secondary, albeit still important. Each of these assessments is, necessarily, subjective in nature. Selecting the "right" niche capabilities will require strategic judgment and imagination.

CHAPTER FOUR

Identifying Candidate Partners

This chapter outlines a conceptual approach to identifying the most promising non-core partners with which to cooperate on niche capabilities. It is organized into two sections. The first section begins with a discussion of the collective action problem outside the confines of a formal alliance relationship. It then explores the possible motivations for partner country participation in coalition operations and highlights opportunities as well as challenges for the U.S. Army in this regard. The second section outlines a framework for selecting candidate niche partner countries, keeping in mind the motivations and challenges identified in the first section. The framework provides a filter for categorizing candidate partner armies according to three criteria: political acceptability, availability for coalition operations, and military capability. This filtering process provides a means for thinking of potential partners in terms of "archetypes" that have one or more key combinations.

Collective Action: From Alliances to Coalitions

This section addresses the challenges of forming and maintaining coalitions of the willing. For political rather than operational reasons, it seems reasonable to expect that coalition operations are the way of the future. Given the likelihood that the international community may question the legitimacy of unilateral action, any such solo operation may have difficulty garnering the support of a large number of governments around the world. And since legitimacy is valuable to policymakers, they will naturally seek to assemble a broad coalition before conducting a military operation. The military implications of this thinking are reflected in the various DoD and Army strategic planning documents, virtually all of which emphasize the development of coalition partner capabilities as a key Interagency, departmental, and service goal.[1]

Despite this, our assumption is that if the United States decides to conduct an operation, it will be carried out regardless of the participation of other partners. This highlights the collective action problem facing the Army and the nation.

There are important differences between alliances and coalitions of the willing. Briefly, alliances are formal arrangements that serve both peacetime purposes of collective deterrence and enhancement of combined military effectiveness through regular multinational training, standardization and interoperability agreements, and the like. Coalitions are temporary, *ad hoc* formations that are typically established for a specific purpose and limited duration. Coalitions are often more politically problematic than alliances because they lack a pre-existing political basis for cooperation. They might be cobbled together without a well-defined strategy, and their members often lack experience working together.

Challenge of Collective Action

In addition to the political challenges implicit in collective action, there are important military challenges. From the perspective of military effectiveness, there is a danger that the involvement of many countries may not necessarily equate to a more capable force. Inexperience in working together, incompatible doctrine and systems, and sheer diffusion of effort often mean that the net military contribution of additional coalition members may be negligible. In extremis, the burden of additional coalition members may actually make the overall coalition less militarily effective. This may then rebound to the political sphere, undermining the political legitimacy of the coalition. Coalitions of the willing are particularly vulnerable to such dynamics.[2]

---

[1]   2004 National Military Strategy (NMS), p. 16.

[2]   Frequent criticism of the coalition assembled for Operation Iraqi Freedom is, arguably, an example of this dynamic. The perception that many members contribute little militarily leads some to doubt the legitimacy of the coalition.

The collective action problem, however, applies to both alliances and coalitions, in terms of what makes countries likely to take part in the first place and the underlying inefficiencies likely to result. Moreover, some coalition members, unhindered by legal, political, economic, or psychological obligation to the other members, may seek additional compensation for their support, and the United States will have to deal with these requests on an individual basis. This is the problem with "outsourcing" in general. The United States may be able to buy increased short-term capacity, but obtaining loyalty to the action and long-term capabilities is more difficult, including guaranteeing that each country will accept the potential costs that come with the risk of collective action.

The challenge of anticipating the dynamics of collective action is a critical one. Several factors must be identified early on, including the core values or goals of the group, as well as whether collective action—rather than individual action or no action at all—will best suit U.S. purposes. To fully understand the collective action dynamics, individual-level dynamics must also be understood. This includes a clear understanding of the individual motivations of each partner country, its anticipated role, the cost and risk it is willing to incur, and the nature of its bilateral relations with the United States and other coalition partners.

Individual-level incentives may curtail the participation of some partners. Their decision not to join a coalition of the willing, or to circumscribe their participation, may involve several factors. First, the nature of the operation might be unpopular domestically, or may become so over time, leading to an early withdrawal of forces. Second, there might be regional political influences that curtail a partner's participation. Pressure from regional hegemons can be a powerful disincentive for smaller states to participate. Third, the partner might not want to divert its scarce military resources to an external operation, particularly if there is some kind of domestic or external threat to contend with.

One significant problem arises when the benefits derived from taking collective action still apply to other partners who did not make a contribution. This is the familiar "free rider" dilemma. Given its enormous military strength relative to any other country, if the United States determines that a particular operation is in its interests, other countries have an incentive to free ride or "easy ride" by making a minimal contribution unless tangible disincentives exist in response to this behavior.

Meeting Challenges in Collective Action

Understanding the collective action problem is the first step in addressing these challenges in a coalition of the willing. It is also useful in explaining intra-alliance behavior, such as in NATO, where the most capable alliance members generally shoulder the military and economic burden. Most allies feel a certain obligation to take part in collective actions approved by the alliance, though this is not always the case. Much depends on the location and nature of the operation, as discussed later in this chapter. Allies sometimes refuse to participate because they have divergent strategic interests that conflict with the proposal to intervene. When an issue falls beyond the range of a given country's immediate geographic influence, its incentive to take action diminishes. The range of a country's concern for international influence is also tied to its level of integration in the international community and volume of international commerce.

As part of the process of developing the selection criteria, it is important to first consider the perceived U.S. and partner political, military, and economic motivations as well as the specific challenges to their participation in coalition operations in a broader context. The next subsection considers the main drivers and challenges for collective action from the U.S. and partner country perspectives.

U.S. Motivations and Challenges of Collective Action

The motivations of both the United States and the partners play a significant role in the decision to establish a long-term security assistance relationship. The U.S. Army would not be well served to invest additional MFC resources in a partner country without first considering the factors surrounding *why* it may want to become a niche partner and understanding how both sides might benefit from the relationship.

Because collective action involves mutual commitment, it is important to consider U.S. and partner motivations to enter into coalitions. The United States has multiple motivations for electing to build and maintain *ad hoc* coalitions of the willing. The most important motivation is political legitimacy. In multilateral operations, the general rule of thumb is that more countries participating will lead to broader political support for taking action. Political legitimacy is strengthened by assembling a coalition of like-minded states, and more so if the coalition includes a diverse group of partners with different political, cultural, economic, and social characteristics. However, political legitimacy can be fragile and temporary. Legitimacy can evaporate quickly if the operation is seen as a failure by the coalition or if governments pull out of it.

The United States may also choose to accept or recruit additional coalition participants to reduce the burden borne by U.S. forces. This must be done carefully, because compatibility issues can undermine the military contribution of some coalition partners. In many cases, however, capable allies and partners can take on a variety of functions within an operation, ranging from combat and stability operations to support functions.

The United States will also face challenges when incorporating coalition partners into multinational operations. The U.S. Army can plan for some of these challenges, but others are harder to anticipate. For example, OIF has demonstrated that operations on the battlefield do not necessarily progress linearly from combat operations to stabilization operations. Therefore, coalition partners must be prepared for a range of operational contexts. National caveats will also be a perennial challenge. These national limitations imposed by each partner on its contingent can be difficult to predict and, in many cases, significantly complicate combined operations.[3]

---

[3] For example, CENTCOM maintains a regularly updated spreadsheet by country by mission area, which is color-coded (red, yellow, and green) according to the missions that coalition partners can and cannot perform in theater. Red means that the country will not perform the mission; green denotes that it will; and yellow reflects uncertainty. Another designation is "yes with permission," meaning that the partner commanders in theater would

Partner Motivations and Challenges for Collective Action

Under what circumstances would a given country wish to participate in a coalition operation, when it has no legal or other obligations to do so? Furthermore, what does it hope to gain from this effort, and what might its national interests or hidden agendas be? There are several possible motivations to consider.

A primary motivation for a partner to join a coalition is that it may expect to reap direct benefits that would be otherwise unattainable. For example, economic incentives can be attractive for less developed countries, and they may explain why countries such as Ghana, Bangladesh, and El Salvador have traditionally proved very willing to provide forces to UN and coalition operations. The desire for international prestige or to bolster regional interests is another potential benefit. A partner might want to either boost its international profile or showcase a particular capability to establish and/or solidify its standing in world or regional politics.

There are other reasons a partner might choose to participate in a coalition operation, even one that is considered domestically contentious. For example, the partner might link its participation to a perceived closer security relationship or even security guarantees from the United States. Other partners with a high degree of uncertainty in their security situation might perceive it to be in their strategic interests to align with the United States in a coalition operation. An extended security relationship with the United States could help them hedge against their regional competitors or threats. For partners that have not yet been able to attain the desired level of cooperation with the United States, participation in a coalition operation may be a way for them to demonstrate worthiness and usefulness.

Other than a few powerful states, such as the United Kingdom, with global security interests, less capable partners without a global power projection capability are more likely to be available for coalition operations regionally and close to their own national borders. Congruity of regional and global interests may sometimes be more important

---

have to seek permission from their capitals before proceeding. Discussions at CENTCOM, January 2005; and with former U.S. commanders in Iraq, February 2005.

than formal alliance arrangements, though as noted earlier, these coalitions may be less predictable.

If the political climate is supportive, some partners may participate in coalition operations to gain real-world operational experience. They might have a particular niche capability that they wish to test or showcase, or they may just want to improve the capabilities of their general purpose forces.[4]

Another motivation to participate that affects the availability factor would be to gain additional training and equipment to improve a particular capability. Some countries capable of contributing specialized forces in high demand, such as special forces capabilities in particular, have benefited from additional U.S.-provided training and equipment. Some of these countries have received increased levels of international military education and training (IMET) and foreign military financing (FMF) security assistance and other economic aid as a result of their participation and support.

Coalition partners must also consider the political, military, and economic challenges that may ensue from taking part in a given operation. In terms of political challenges, there are several factors to consider. First, if the operation is unpopular at home or becomes unpopular over time, the challenges for a government are substantial. As we have seen regarding Iraq, coalition military operations can become a key election issue.[5]

There are also possible political consequences of aligning, even if only temporarily, with the United States, if the country is also tied economically, politically, and culturally to a potential U.S. adversary. Russia and China, for example, have attempted to dissuade some Central Asian partners from granting U.S. access to military facilities in the region.[6]

---

[4]   Discussions with U.S. liaison officers embedded in the Multinational Division-Center South in Iraq (working with the Latin American countries), April 2005.

[5]   For example, Spain, Bulgaria, Ukraine, and Japan are all examples of how coalition operations unsupported by the populace influenced election results, or required candidates to make campaign promises (i.e., force drawdown if elected).

[6]   In July 2005, the Russia/China-dominated Shanghai Cooperation Organization, which also includes the Central Asian states as members, issued a joint communiqué urging the United States to set timetables for force withdrawal from their territories.

A potential coalition partner must also consider the military challenges associated with coalition operations. The unpredictable nature of military operations, along with mishaps stemming from lack of preparation and interoperability deficiencies, can result in casualties. At present, the practice of evaluating coalition partner capabilities prior to deployment to theater is *ad hoc.* Thus, the United States and key allies have incomplete information on the military capabilities of coalition partners that have less experience working in an operational environment. Moreover, the complicated and cumbersome command and control system in a large multilateral operation brings inherent military challenges and can be confusing to partner countries.[7] Another military challenge to the partner is the temporary decrease in available military capacity, since whatever is deployed with the coalition cannot be employed domestically. This is a particularly acute problem if a partner country's engineering, aviation, or policing capabilities, for example, suddenly become needed to deal with a domestic emergency.

Economic challenges are no less important to consider from the partner's perspective. A smaller, less economically robust partner country could incur significant operational costs. The costs will vary in terms of the number of forces deployed and duration of participation. However, the impact on a national economy could be quite substantial.

Framework for Selecting Candidate Niche Partner Countries

The Niche Capability Planning Framework focuses Army MFC resources on non-core partner armies that have the potential to be niche partners. To select the best possible set of niche partners, it is necessary to separate promising partners from the rest. Our proposed framework for making these judgments categorizes potential partners according to three criteria: the political acceptability of the potential partner, its availability for coalition operations, and its level of military

---

[7]   Discussions in Kyiv, Ukraine, with former Ukrainian commanders deployed to Iraq, February 2005.

capability. The framework suggests which combinations of these three criteria make for the most promising niche partners. It also outlines a portfolio approach to selecting niche partners with a range of political, social, and geographic characteristics in order to minimize political risk.

A few caveats must be highlighted up front. First, it is our assumption that analyzing the political appropriateness of potential partners is inherently a subjective process, and there is no set of criteria or a methodology that will yield an objectively verifiable optimum set of potential partners. Our framework uses indicators that are both subjectively contestable and variable over time. Nevertheless, we believe it will help Army planners identify the key factors in selecting niche partners and, as importantly, structure intra-Army discussions regarding MFC planning and the ASCS.

Second, it is not possible to accurately predict which countries will participate in any given coalition operation with the United States. A number of factors will influence this decision, including the partner's political interests at the time of the operation and the type of contingency envisioned. The Army therefore incurs some degree of political risk with each niche partner, measured as the likelihood that the partner will opt not to join a future coalition.[8] This political risk is to some degree unavoidable. The purpose of this framework is to minimize that risk, but it cannot be altogether eliminated.

In developing criteria for assessing potential niche partners, the study team first considered selection criteria based along the lines of the broadly accepted "DIME" (diplomatic, informational, military, economic) scheme.[9] This effort quickly became unwieldy because of

---

[8] Thomas Szayna, Frances Lussier, Krista Magras, Olga Oliker, Michele Zanini, and Robert Howe, *Improving Army Planning for Future Multinational Coalition Operations*, Santa Monica, CA: RAND Corporation, MR-1291-A, 2001, pp. 37–38.

[9] Previous RAND work attempted to identify promising candidate partners for coalition operations around the world. In *Improving Army Planning for Future Multinational Coalition Operations*, Szayna et al. developed the Military Compatibility Assessment Tool (MCAT). This adaptation of the RAND-developed DynaRank Excel-based tool aims to help Army planners identify compatibility with the U.S. Army in broad battlefield functional areas, such as maneuver or fire support.

the virtually unlimited number of factors that could be placed in any of these categories. In addition, it was nearly impossible to rank order the factors or compare across factors, because the approach produced too many overlapping indicators relative to the DIME categories. We therefore determined that what was needed was a parsimonious filtering process based on just a few key factors. Those factors could then be fleshed out in greater detail and could serve as a heuristic device for planners. The three key factors selected are *acceptability, availability,* and *capability.*

Acceptability and availability derive directly from the challenges of collective action. Acceptability is a measure of the willingness of the United States to accept the potential partner as a niche partner and likely coalition member. Availability measures the potential partner's willingness to become a niche partner with the United States and to participate in future coalition operations.

Figure 4.1 provides an overview of a framework for selecting niche partners. The three circles of the Venn diagram are not weighted evenly. The most important factor is the acceptability of the partner to the U.S. government. The second most important factor is the willingness of the partner to participate as a niche partner. This factor is ranked second because, while important, it potentially can be influenced by U.S. diplomatic efforts and security cooperation activities. The capability criterion ranks third, largely because the overall Niche Capability Planning Framework is designed to provide niche capabilities to partner armies. If the partner already possesses a well-developed niche capability, that is an added benefit but is not a discriminating factor at the outset.

Acceptability. The acceptability factor is a political "litmus test" to determine whether or not a partner is eligible to receive MFC resources from the United States. Determining a partner's overall acceptability can also help the Army to identify potential impediments to the cultivation of niche capabilities. It is a structured means of asking whether, given the current administration's strategy and policies, developing a closer security cooperation relationship with the partner country is politically acceptable. Further, it is important to consider where a particular

Figure 4.1
Factors to Consider

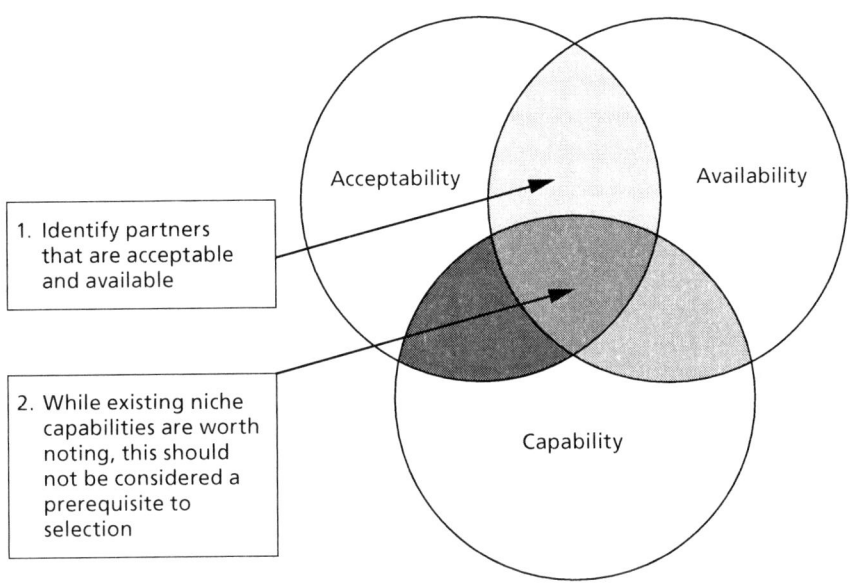

1. Identify partners that are acceptable and available

2. While existing niche capabilities are worth noting, this should not be considered a prerequisite to selection

Acceptability

Availability

Capability

RAND *MG563-4.1*

partner falls in terms of priority within the context of the ASCS and the OSD Security Cooperation Guidance and the COCOM's TSCP.

We identified two indicators to help the Army determine acceptability. These are: *common political values* (shared among the United States and the partner) and *diplomatic relations*. First, in terms of common political values, it should be determined whether there are processes in place in the partner that lend themselves to democratic practices. One example could be the presence of a functioning and fair legal system, which could indicate an established basis for mutual understanding with the United States and the partner country. However, the acceptability factor should not be too limiting in the candidate partner selection process. The partner's regime and government structures need not mirror the United States in terms of having a fully functioning democratic system and a market economy.

Second, in terms of diplomatic relations, the partner's receptivity to discussions on key issues (e.g., nonproliferation) should be considered, as well as the level and nature of bilateral exchanges, such as regular meetings at multiple levels (presidential, ministerial, working, etc.), which can indicate a basis for shared political views. Signed and ratified military agreements, such as the bilateral Status of Forces Agreement (SOFA) and the Acquisition and Cross Servicing Agreement (ACSA), are also important to consider. The conclusion of these agreements can indicate a willingness on the part of the partner country to deepen military cooperation with the United States. Further, if a partner has signed what is considered to be a contentious agreement, such as agreeing not to render U.S. service members to the International Criminal Court, this may be an indictor of the partner's willingness to work closely with the United States.[10] The signing and ratifying of such agreements may demonstrate a higher degree of commitment to deepening political and military cooperation.

Availability. After determining a partner's level of acceptability to the United States, the next step would be to assess the country's availability for coalition operations. From the U.S. perspective of the partner's situation, the question is whether or not the partner is willing to develop closer relations with the United States to cultivate deployable niche capabilities, and what evidence points to this conclusion. The following factors can be considered as a starting point. First of all, the *partner's level of interest* should be determined, assessed in part by its desire to address and influence regional and/or global security problems. Is the country looking to increase its international prestige? Does it seek economic gain? Does it simply want to be part of the larger war on terrorism? Does it want to improve its status in a particular region? This factor is closely linked to the partner's *proximity* to regional hot spots and unstable regions as an indicator of its likely willingness to participate in a regional operation. Generally speaking, for less capable partners there is a greater impetus to get involved if the operation is in the same region.

---

[10] The State Department currently requires such an agreement, or a presidential waiver, to provide Title 22 security assistance to a foreign state.

Second, the partner's *political, legal, and economic situation* relative to the United States must be considered. Regime stability in the prospective candidate country is important to consider, especially if the United States is considering either allocating significant resources toward building or augmenting a niche capability, or supporting the cultivation of a more sensitive capability, such as civil disturbance training, in that country. From a political perspective, the orientation of the current regime, popular opinion of the United States, and the role of the political opposition are also important indicators to monitor.

The ability of a third country or organization to influence the foreign policy of the potential partner should also be considered. This influence might run counter to U.S. interests.

Another valuable measure is past participation in U.S.-led coalitions. Some countries have accorded "soft support" such as intelligence/information sharing, overflight rights, and logistical support in recent operations. Some partners have provided more tangible support, exemplified by the deployment of large contingents to hostile environments in Iraq and Afghanistan. Other factors to consider include number and type of forces deployed, whether the partner was in a lead or supporting role, and the number of rotations completed.

The potential partner's security cooperation relationship with the United States is another indicator of availability. A dynamic security cooperation relationship could be an indicator of a lack of barriers to military cooperation. For recipients of U.S. security assistance (i.e., supported partners), Army planners might consider the following questions: How enthusiastic are they about working with the U.S. military? Do they show up for events and actively participate in the planning process for them? Do they make requests for specific types of activities, or are they simply in receive-mode for U.S. ideas? Do they participate in both bilateral and multilateral (regional) exercises? Do the respective ministries of defense send promising commissioned and noncommissioned officers to the United States for IMET and other training, or do they send intelligence officers? All of these factors are telling of the overall perspective in which the respective defense establishments view their relationship with the U.S. military.

In terms of legal issues, one issue to consider is whether or not there are legal impediments to deploying the partner's forces outside the national borders. For example, for countries that have conscript forces, is it permissible to deploy those forces outside the home territory?

Finally, on the economic side, the United States might consider the partner's anticipated ability to invest in the building and sustaining of niche capabilities using its own national resources. Broadly speaking, factors such as defense budget relative to GNP, and general economic trends (growing, holding steady, or shrinking) relative to their ability to support their deployments may be good indicators of availability. Moreover, it is important to consider whether supporting the development of a particular capability may have adverse effects on defense reform efforts, especially if the partner simply cannot afford to do both.

Capability. The final step of the process is to consider the amount of assistance required to build an interoperable, sustainable niche capability. Simply stated, does the partner have an appropriate defense structure in place to support the cultivation of deployable niche capabilities? How much training and equipment is required? Army planners can evaluate the capability factor relative to the following indicators: leadership, personnel, equipment, deployability, sustainability, finance, and existence of nascent or advanced niche capabilities.

In terms of *leadership*, three factors should be considered. First, whether the military is subordinate to a civilian authority in theory and in practice, as indicated by civilian policy structures with ultimate authority in military affairs. Second, whether the partner has instituted a functioning chain of command extending from tactical to strategic echelons and terminating with civilian leaders. Third, whether the partner has established procedures and processes for exerting command of deployed forces.

The capabilities of the military *personnel* should also be taken into account. First, whether the partner operates a system of formal professional military education for commissioned and noncommissioned officers. Second, whether the professional military curriculum covers the key topics required, including the principles of multilateral military operations, international law, operational law, and the areas of functional and staff expertise required to manage the niche capability in question. Third, the level of international experience of the force should be taken into account as an indicator of precedence for participating in out-of-area operations.

Regarding *equipment*, the key indicator is the sophistication and maintenance status of the equipment currently owned by the partner. This is a broad measure of the partner's ability to sustain and maintain the equipment required for a niche capability. The size and stability of the partner's procurement budget is also a useful measure in this regard.

A third factor under the capability heading is *deployability and sustainability* of the partner's capabilities. The partner's assets, specifically infrastructure at home and combat support/combat service support for operations, are an indicator of its ability to deploy and sustain the niche. Also, having the necessary processes in place (logistics management system, tooth-to-tail support capability, etc.) is an indicator of a partner's ability to sustain the niche.

Trends in public finance, most importantly the defense budget, are important insofar as the strength and orientation of a partner country's public finances says much about its ability to perform the full range of tasks required as a force provider, including the development, maintenance, and sustainment of the niche without relying entirely on U.S. support. Within this context, if it is possible to discern a country's ability and willingness to spend its own national resources on particular niche capabilities, that would be very useful information in the selection process.

Finally, the partner may already have *niche capabilities* that could augment U.S. capabilities or fill existing gaps. The Army might consider the type of capability, and to what extent the partner has allocated its own national resources toward its development or sustainment. The last step in selecting candidate partners is the filtering process, as shown in Figure 4.2.

Figure 4.2
The Filtering Process

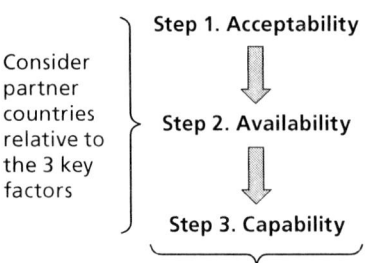

Step 1. Acceptability

Consider partner countries relative to the 3 key factors

Step 2. Availability

Step 3. Capability

1. Justification: Cannot provide assistance to partners to develop niche capabilities if politically unacceptable to U.S.

2. Justification: Once deemed politically acceptable, consider availability from partner's perspective

3. Justification: Once deemed acceptable and available, consider capability of assets partner wishes to develop and potentially deploy

Illustrative list of candidate partners

RAND *MG563-4.2*

Partner Archetypes

We next used the filtering process to parse potential partners into "archetypes" that have one or more key combinations of acceptability, availability, and capability. Although many different archetypes could be developed from these combinations, two in particular are most relevant for the Niche Capability Planning Framework. Countries in the first archetype, termed *emerging allies and partners*, are generally supportive of U.S. military operations but are held back by a lack of necessary capabilities and resources. This is one of the better groups to focus upon for niche capabilities cultivation precisely because these countries are politically acceptable and available but require some assistance in developing their capabilities for coalition operations. Also, they may be more willing to take some direction from the United States in terms of what types of capabilities to develop.

Through their involvement in coalition operations, these countries gain valuable operational experience and improve their overall compatibility with the United States. But because their capabilities are less developed, their participation may increase the overall risk to the coalition. These risks can be diminished through targeted Army MFC activities. Overall, countries that fit this description are likely to be good candidates for niche capabilities cultivation.

The second archetype, termed *potentials,* includes partners that typically demonstrate a reluctant willingness to work more closely with the United States in a coalition environment. Some have contributed forces in the past, and some have indicated that future deployments could be possible. These countries are politically acceptable, may or may not be available (i.e., contributing forces to U.S.-led coalition operations for any length of time would be difficult for them), and need some help developing their deployable capabilities.

Focusing U.S. Army security cooperation on these countries would allow for the development of a broader base of potentially capable coalition partners. They would have the opportunity to gain valuable operational experience and improve their overall interoperability with the United States. However, if the capabilities the partner wished to deploy are not compatible, this could adversely affect the operational effectiveness of the coalition and lead to unnecessary casualties. Moreover, if the domestic political environment of the partner is fluctuating, this could lead to early withdrawal of forces. Still, the U.S. government and particularly DoD might be able to change the availability factor of the partner over time, altering its strategic calculus, through security cooperation activities targeted at various levels. Therefore, there are grounds to believe that the potentials are good candidates for niche capabilities cultivation.

Because countries within these categories are politically acceptable to the United States, the *emerging allies and partners* and the *potentials* archetypes are already or have the promise of being available to deploy forces for coalition operations, and simply need some assistance developing their niche capabilities. Therefore, the Army might consider how it can better focus its capabilities-building security cooperation resources on those partners that fit this description.

Additional Criteria to Consider to Manage Risk

In addition to considering the acceptability, availability, and capability of individual potential partners, it is also advisable to keep in view the aggregate profile of MFC investments in niche capabilities. As noted above, niche partnerships will create some degree of political risk for the United States. One major concern in this regard is the possibility that the Army will devote considerable MFC resources to cultivating a niche capability in a partner's army, only to see that partner refuse to participate in subsequent coalition operations. This would be especially problematic in cases where the Army had accepted additional risk in its own force planning on the understanding that a niche partner would provide the needed capability.

To manage this type of political risk, Army planners should view niche MFC partnerships as a portfolio of investments in capabilities. Niche capabilities should be viewed as insurance against large, prolonged, or specialized requirements. Furthermore, Army planners should take care to maintain as much geographic and political diversity as possible in niche partnerships.

Political Diversity. Each niche partner government will possess its own national interests and political preferences. Despite a general willingness to cooperate with the United States on developing a niche capability, governments will disagree with some U.S. policies and will have their own political reasons for joining, or refusing to join, a particular coalition. At times, a partner may refuse a U.S. request for support because the U.S. intervention may be unpopular among the niche partner's population, it may be unwilling to become involved in certain types of conflicts, or there may be regions of the world in which a partner feels uncomfortable operating (former colonial powers are an example). Even in cases where a partner's participation is not precluded outright, political considerations may shape its participation in important ways. Ongoing operations demonstrate that in the context of a coalition of the willing, partner countries typically have a clear idea of the types of missions they would like to perform in a given operation, and they tend to stick to those national preferences and caveats. In Iraq, many partners are considered risk averse by U.S. standards, preferring to take part only in noncombat operations such as policing, reconstruction, and humanitarian assistance. Others prefer to operate where the environment is peaceful or benign.

Given this political risk, it would therefore be wise for Army planners to consider the political, social, and cultural profile of its niche partners. For any given niche capability, the Army will want to develop multiple partnerships across a variety of political, social, and cultural characteristics. As an example, the Army would not want to develop a niche capability in a group of partners that all refused to conduct combat operations, or were all of the same ethnic and cultural background, or all the same political persuasion.

Geographic Distribution. It would also be wise for the Army to consider geographic distribution in the process of selecting candidate countries for niche capabilities. The analysis in this study suggests that many desirable niche capabilities are likely to be concentrated in Eastern Europe, Latin America, and Southeast Asia. However, future operations may be conducted in any region of the world. By nature, coalition composition is dependent on many factors, and it may not be a good idea to count on a specific group of countries in a specific region to be available for coalitions of the willing, regardless of mission type. Niche partners in a region where the United States is conducting an operation will also bring critical expertise that cannot be replicated by a niche partner from another region. The Army should therefore consider geographic distribution as it selects its niche partners.

Conclusion: An Integrated Approach to Identifying Candidate Partners

This chapter outlines a planning framework to assist Army planners in selecting niche partners. The process of selecting niche partners is inherently subjective and complex. In our assessment, a selection process that focuses on the acceptability, availability, and capability of potential partners, keeping in view the larger picture of political and geographic diversity, can serve as a useful heuristic device for Army planners.

CHAPTER FIVE

Cultivating Niche Capabilities

This chapter describes a conceptual approach to cultivate niche capabilities in non-core partner armies. It begins by discussing the process in general terms, and then outlines an approach that parallels the Army's own Title 10 force development processes for organizing, equipping, and training U.S. Army units. This chapter concludes by discussing some of the challenges of cultivating niche capabilities.

Niche Cultivation Approach

Cultivating niche capabilities in non-core partner armies might seem an unusual mission for the U.S. Army. In fact, however, organizing, training, and equipping formations is the core of the Army's Title 10 mission. A large portion of the U.S. Army, often labeled the Institutional Army, exists to organize, train, and equip the Army to meet the needs of the joint force commander.[1] A significant number of these organizations, ranging from Headquarters Department of the Army to TRADOC schoolhouses and Army Materiel Command laboratories, routinely participate in security assistance and Army security cooperation activities. A great deal of the expertise required to cultivate niche capabilities in foreign armies therefore already resides in the Army.

---

[1] The Institutional Army is often referred to as the "TDA Army" after the planning documents that authorize these organizations (Table of Distribution and Allowances). The Operational Army is often referred to as the "TOE" Army for analogous reasons (Table of Organization and Equipment).

Conceptually, there are two ways to approach the task of cultivating niche capabilities in non-core partner armies. One approach is to attempt to replicate, to the extent possible, the organization, training, and equipment of U.S. Army formations in these partner armies. This approach promises to be the easiest to implement because the U.S. Army organizations and personnel responsible for cultivating these niche capabilities are already familiar with the U.S. Army formations. It would also result in the greatest degree of compatibility between U.S. Army and the new niche units, because U.S. Army units are already familiar with the requirements of working with analogous U.S. units. However, care must be taken to avoid producing resentment in partner armies, a possibility if the U.S. Army is seen to be simply trying to replicate itself in the partner's Army. Furthermore, an American-pattern unit may not be easily integrated into the partner's overall defense structure, making it more difficult to sustain.

A second approach would be to make the minimum necessary adjustments to an existing partner army formation in order to allow it to conduct its new mission. This would have the advantage of changing the unit as little as possible, avoiding the disruption and possible resentment of creating a new model formation and allowing the partner to maintain the uniformity of its armed forces to the maximum extent possible. It would have the disadvantage of being more difficult for the U.S. Army to implement, as every niche partner would have a slightly different niche unit design, and the niche units would be less compatible with U.S. forces due to their unique designs.

The Niche Capability Planning Framework melds elements of these two approaches, but in practice it reflects the first more than the second. The demands of compatibility, ease of implementation, and familiarity to U.S. Army officers require that niche units be as similar as possible to the U.S. Army units with which they will operate. Therefore, the approach for niche cultivation outlined below describes an approach to creating niche units that are as interchangeable as possible with U.S. Army units. However, we recognize that in certain circumstances political factors will require that modifications be made to this approach, and we have outlined in each step how this might be accomplished without sacrificing the compatibility benefits of the niche approach.

Cultivation Approach

Figure 5.1 depicts the proposed approach for cultivation of niche capabilities. It clearly shows that the organize, train, and equip portions of the Army's traditional Title 10 mission are most relevant to the niche capability strategy. It further depicts how the organization and training for a niche unit can be derived largely from the Combined Arms Training System (CATS) documents for analogous U.S. Army units. It also illustrates how the equipment requirements for niche units can be derived from the Modified Table of Organization and Equipment (MTOE) data available for analogous U.S. Army units. In most cases, Army planners will wish to modify the niche unit organization, training, and equipment from the U.S. original. Nevertheless, existing processes and documents represent a valuable template for niche cultivation.

Figure 5.1
Conceptual Approach for Cultivation of Niche Capabilities

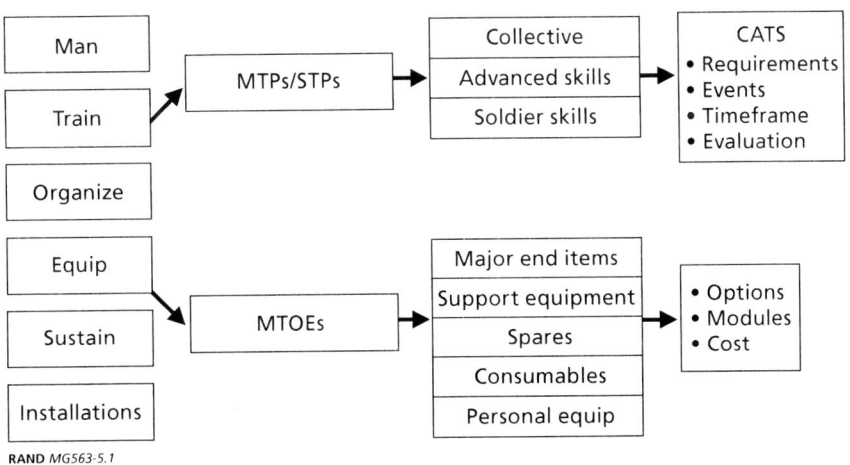

RAND *MG563-5.1*

Organizing Niche Units[2]

In analyzing how to organize non-core partner niche capability units, Army planners can look to the Army's own force development processes for guidance. Indeed, as noted, in most cases it will be best if foreign niche units are organized as closely as possible to an analogous U.S. Army unit. While this may cause some discomfort in some foreign armies, it will greatly increase compatibility in the field with U.S. Army units, and it will simplify the Army's task of providing institutional and field support to the niche unit.[3] However, it should be noted that it is not imperative that partner units be the same size as their U.S. counterparts; indeed, many U.S. partner armies have very small units, so matching U.S. TOEs in terms of size would be inappropriate.

In most cases, Army planners will want to modify an existing U.S. Army TOE to create the niche unit. This would involve selecting a current U.S. Army TOE (or MTOE) for an existing unit that most closely matches the mission and desired capability of the niche unit. For example, if the U.S. Army desires to create light infantry battalions in niche partner armies, Army planners might use the most recent Infantry Battalion MTOE published by TRADOC.[4]

Army MTOE documents provide a wealth of detailed information about operational formations. They define the unit's structure, organization, and subunits down to the lowest level of detail. They define exactly how many personnel, of which specialties and ranks, will fill each unit and subunit. They define exactly how the unit should be

---

[2] In this chapter we refer to "niche units." When we use this term we refer to units in non-core partner armies that possess the niche capabilities being cultivated by the U.S. Army.

[3] Some foreign armies may feel that they are being "Americanized" to an unacceptable degree if niche units in their force structure are organized and trained in a manner that is identical, or nearly so, to U.S. Army units. Others may feel comfortable, or even perceive some prestige value in being organized and equipped in this manner. Whenever the Army planners meet resistance along these lines, they may wish to modify the niche units to bear a less direct resemblance to U.S. units, work to assuage partner concerns, or simply accept these concerns. Alternatively, Army planners might consider such concerns a warning sign that the long-term relationship required by the niche capabilities approach may not be viable with armies that would allow such concerns to affect cooperation.

[4] "Infantry Battalion (Infantry BCT)—MTOE," Document Number 07415GNG12, dated 24 June 2005.

equipped (on which more later). As such, they provide Army planners with an excellent template for developing niche unit designs. In most cases, it should be unnecessary to create an organizational design for a niche unit from scratch.

Army planners will in most cases, however, not wish to simply replicate U.S. units within foreign armies. Instead, they will wish to tailor the capabilities, and therefore the structure, of foreign niche units to match shortfalls within the Army's capabilities. For instance, to use the light infantry battalion example, Army planners might wish the foreign niche light infantry battalion to have less indirect firepower, but more civil affairs capability, than standard U.S. light infantry battalions. They might therefore choose to modify the U.S. Army MTOE, as applied to the foreign niche battalions, to delete the mortar platoon and add a civil affairs platoon. These changes could be easily reflected in a modified TOE developed by Army planners that substitutes data from a civil affairs platoon MTOE for the mortar platoon assets in the infantry battalion MTOE.[5] In this manner, existing Army MTOEs can be utilized to modify and tailor foreign niche units while maintaining essential compatibility and familiarity with U.S. Army units.

Once an MTOE is available for the proposed niche unit, it will provide Army planners with a comprehensive understanding of the structure of the unit, the number and types of personnel required to fill out the unit, and the unit's basic capabilities. Moreover, by using U.S. MTOEs as the basic template, Army commanders in the field who will work with the niche unit should understand intuitively the potential capabilities it offers as well as the compatibility challenges likely to arise. Using U.S. Army MTOEs therefore promises to simplify the planning challenge of developing niche units and the operational challenge of incorporating them into coalition operations.

Training Niche Units

Once Army planners have established an organizational design for the niche unit, they must then determine how much training, of what

---

[5]   Light infantry is not a candidate niche capability we recommend. It is provided here solely as an example to illustrate the conceptual framework.

types, is needed to create the required capabilities within the niche unit. Again, Army planners will benefit from using documents already in place for analogous U.S. Army units. The Army's basic approach to training is to build from individual soldier tasks, to small unit collective tasks, to large unit collective tasks. Within each level of training, the Army adopts a "crawl-walk-run" approach that starts with the very basics of a task and proceeds in a deliberate manner to more complex aspects of the task. Figure 5.2 outlines this approach.

Army planners can begin by creating a Mission Essential Task List (METL) for the niche unit, likely based on an analogous U.S. unit. For our light infantry battalion example, a METL could be constructed based on Field Manual (FM) 7-20, *The Infantry Battalion,* and Army Train and Evaluation Program (ARTEP) 7-20-MTP, *Mission Training Plan for the Infantry Battalion.*[6] FM 7-20, like all Army doctrinal manuals, outlines the primary missions and tasks to be undertaken by the infantry battalion. It is therefore an effective, if broad, guide for Army planners establishing an analogous niche unit. ARTEP 7-20 provides much more detail, specifying 43 mission essential tasks

Figure 5.2
Army Approach to Training

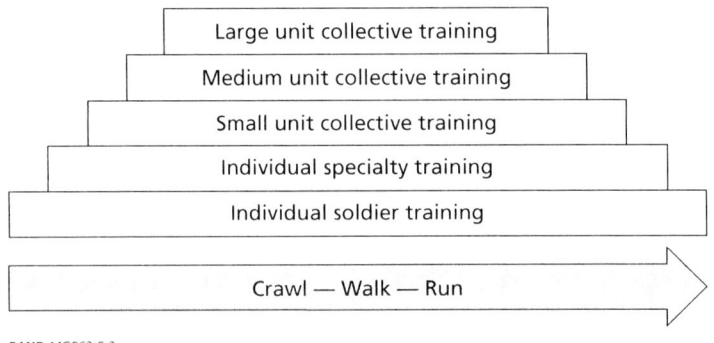

RAND MG563-5.2

---

[6] Field Manual 7-20, *The Infantry Battalion,* Fort Monroe, VA: U.S. Army Training and Doctrine Command, 1992; Army Training and Evaluation Program, 7-20-MTP, *Mission Training Plan for the Infantry Battalion,* Fort Monroe, VA: U.S. Army Training and Doctrine Command, 2001.

at the light infantry battalion level. More than 30 additional mission essential tasks at the company and lower levels are outlined in ARTEP 7-10, *Mission Training Plan for the Infantry Rifle Company*. These 75 tasks provide a very detailed outline of the tasks the Army expects its light infantry battalions to execute.

As with the organizational design, Army planners will in most cases not wish to simply replicate the METLs for analogous U.S. Army units. Rather, they will probably wish to tailor the METL of niche units to the shortfalls in U.S. Army capability. Following the light infantry battalion example, Army planners might wish to remove the METL tasks associated with the battalion's mortar platoon and add METL tasks for a civil affairs platoon. These civil affairs tasks might be directly imported from the doctrine and ARTEP materials for the civil affairs branch, or they might be developed from scratch. Such changes could be reflected in a modified Mission Training Plan (MTP) published for the niche unit.

Once Army planners have defined the organization, mission, and tasks for the niche unit, they can begin to determine the amount and types of training required to develop those skills. Again, the Army's own internal planning process provides a useful resource. The CATS of training plans provides a comprehensive training regimen for each of the collective tasks assigned to U.S. units (see Figure 5.3). This can provide the basis for Army planners to determine how many training events will be required to bring a niche unit to full capability. For example, ARTEP 7-10, *Mission Training Plan for the Infantry Rifle Company,* specifies that to prepare a light infantry company to conduct an attack, company leaders should conduct one three-hour class, one four-hour Tactical Exercise Without Troops (TEWT), and one twelve-hour force-on-force Situational Training Exercise (STX) using the Multiple Integrated Laser Engagement System (MILES). According to the MTP, after completing these events to standard, the company is qualified to conduct this task. The MTP contains similar guidance for the other 31 tasks specified for the light infantry company. Importantly, the CATS documents also give guidance to Army planners on how to evaluate units conducting each training event for each task. For example, it outlines how evaluators can measure whether a light infantry company is

Figure 5.3
Example of Training Literature for Light Infantry Battalion

ARTEP 7-20 *MTP for the Infantry Rifle BN*

ARTEP 7-10 *Mission Training Plan for the Infantry Company*

ARTEP 7-8 Drill *Battle Drills for the Infantry Platoon and Squad*
ARTEP 7-92 *MTP for the Recon Platoon and Squad*

STP 7-11B1-SM-TG
Soldier's Manual and Trainer's Guide, MOS 11B, Infantry, Level 1
STP 7-11B24-SM-TG
Soldier's Manual and Trainer's Guide, MOS 11B, Infantry, Level 2, 3, 4

STP 21-1-SMCT *Soldier's Manual of Common Tasks, Skill Level 1*
STP 21-24-SMCT *Soldier's Manual of Common Tasks, Skill Level 2, 3, 4*

FM 7-9 *The Infantry Rifle Platoon and Squad*
FM 7-10 *The Infantry Rifle Company*
FM 7-20 *The Infantry Battalion*

RAND *MG563-5.3*

successfully conducting its Company Attack STX, providing multiple go/no-go indicators for each task assigned to the unit.

The CATS documents supply a useful template for building training plans for niche units. Based on the missions and METLs selected for the niche unit, Army planners can use the CATS documents to define how many training events should be arranged for the niche unit, of which types, and in what order. As in other areas, Army planners are likely to opt to modify the CATS documents to better suit the niche units, substituting training events that are more feasible overseas for CATS-specified events that call for training capacity unlikely to be available in the partner nation (such as MILES equipment and/or urban training facilities).

These collective training requirements can also be augmented, if necessary, by individual skills training requirements as outlined in Soldier Training Plans. These documents outline the individual tasks that every soldier within a specialty should be able to accomplish before the unit as a whole undertakes collective training. In terms of our light infantry battalion niche unit example, Army planners could consult existing documents that outline precisely what each infantry soldier should be capable of doing, as an individual, to support the unit's collective tasks.[7] Together with the CATS data for collective training, these individual soldier training documents give Army planners a comprehensive outline of the likely training requirements for a niche unit.

Equipping Niche Units

The concomitant step will be to determine how the niche unit should be equipped. Here again, existing U.S. Army processes and data can be very useful to Army planners. In particular, the MTOEs developed for each U.S. Army unit specify how the units should be equipped to fulfill their assigned tasks. This equipment data is provided down to the individual item and covers everything from major end-items such as vehicles to uniforms and personal equipment. Where a niche unit has been organized according to an existing U.S. Army MTOE, and trained according to that unit's CATS documentation, Army planners will be able to identify much of the unit's key equipment by consulting the U.S. Army unit's MTOE.

However, as with organization and training, Army planners will want to modify the MTOE for the niche unit. There are several reasons for this. First, any modifications that have been made to the unit's mission, organization, and training would of course be reflected in equipment as well. For instance, following our light infantry battalion from above, if the niche light infantry battalions have been stripped of their mortar platoons, then there is no need to provide the unit with the mortars, trailers/vehicles, communications, and other paraphernalia that accompany the mortar platoon. Likewise, if a civil affairs platoon has been added to the niche light infantry battalion, then appropriate equipment will need to be added. A basic template for this equipment is available in the form of MTOEs for U.S. Army civil affairs platoons.

Additionally, many U.S. military items are not available for export under some circumstances. For example, DoD has a longstanding

---

[7] For example, STP 7-11B1-SM-TG, *Soldier's Manual and Trainer's Guide*, MOS 11B, *Infantry Skill Level 1*, and STP 7-11B24-SM-TG, *Soldier's Manual and Trainer's Guide*, MOS 11B, *Infantry Skill Level 2, 3, and 4*.

reluctance to export advanced night vision devices to even close friends and allies. Providing them to niche partners is likely to be problematic. Likewise, there are obviously concerns that communications and computer equipment given to partners might be compromised, providing a communications intelligence advantage to future adversaries.

Finally, many items of U.S. Army equipment are not suitable for some partner armies. U.S. soldiers are technically savvy and possess a strong maintenance culture. Sophisticated or delicate equipment that might be present on U.S. Army MTOEs might not be appropriate for niche unit MTOEs.

Bearing in mind all of these caveats, U.S. Army MTOEs still provide an excellent template and source of data for analyzing the equipment requirements of niche units. Moreover, providing niche units with standard issue U.S. equipment in prescribed numbers will also have important implications for compatibility. The physical aspects of interoperability would be greatly simplified by common equipment. Likewise, the conceptual aspects of logistical and support planning will also be simplified if niche units have standard equipment in standard allocations. The provision of U.S. equipment in this way can be costly, but the tradeoffs of deploying coalition forces with incompatible equipment (i.e., risks to U.S. and coalition forces and the local population) would likely outweigh those costs. Overall, it is more a matter of channeling U.S. security assistance (e.g., Foreign Military Financing and Foreign Military Sales, and Excess Defense Articles grants, etc.) toward the purchasing of the most appropriate, compatible equipment than of seeking new funding streams within Army security cooperation activities.

Key Challenges

While the Army's approach to organizing, training, and equipping its own formations provides a useful template for cultivating niche capabilities, this process will not be without its challenges. Our assessment suggests that there are five significant challenges that may confront Army planners in the cultivation phase. First, as already noted, it will be difficult to design a niche unit that meshes smoothly with the U.S. Army without making it so alien to the partner army that it becomes politically and administratively difficult to sustain. Army planners will be obliged to strike a careful balance with each partner in this regard, separating those characteristics of the niche unit that are critical to its compatibility with U.S. units from those that might be sacrificed to create a sense of ownership and familiarity in the partner army.

The second major challenge will be ensuring that the niche unit receives the training it requires over the long term. In the near term, the Army has an adequate capability to train niche units to standard. However, maintaining this level of proficiency will require periodic training that is both onerous and expensive. The Army cannot afford to manage the steady-state training cycles of its foreign partners, but Army planners will need to develop some means of monitoring partner training and providing resources and expertise to support this training where required.

The third major challenge will be human rights vetting. The State Department routinely examines the rosters of foreign units receiving U.S. training to ensure that there are no reported human rights violators present. While this process is stringent, it is not foolproof, and Army planners will want to make doubly sure that individuals or units with unseemly reputations do not take part in niche training. Furthermore, Army planners must ensure that the skills and techniques taught to partner units meet the highest standards of human rights and the laws of war.

The fourth major challenge will be sustainment of the equipment provided to the niche unit. The Army should, in most cases, have little trouble delivering serviceable equipment to partner armies along with support packages containing the tools, spares, and consumables required to maintain the equipment. However, over the long run the Army will likely require the partner to provide sustainment support for the niche unit. This has been a key weakness of U.S. security assistance for decades, and Army planners will want to devise a monitoring and support process to ensure that niche unit equipment is ready when needed.

The fifth major challenge will be identifying and coordinating the appropriate mix of activities, DoD resources and programs, and Interagency resources to effectively and efficiently cultivate niche capabilities. This process is the focus of Chapter Six.

Conclusions

This chapter outlines a conceptual approach to cultivate niche capabilities in non-core partner armies. The approach parallels the Army's internal Title 10 force development processes. Though this could make some potential partner armies uncomfortable, paralleling the Army's own processes has several advantages. First, it is relatively simple and straightforward for the U.S. Army to plan and implement, as it mirrors the Army's own day-to-day force development processes. Second, it promises to produce niche units that are as compatible as possible with U.S. Army units, by dint of their organization, training, and equipment allocations.

The process described in this chapter represents only a single step in the Niche Capability Planning Framework. In identifying niche unit organization, training, and equipment requirements in such detail, Army planners set the stage for the next step of the process, which is identifying how the Army will provide the required advice, training events, and equipment to potential niche partner armies. In essence, developing modified MTOEs and CATS documents for niche units will allow Army International Activity planners to design a synchronized program of activities to cultivate the required capabilities within the niche units. The next chapter outlines this process.

CHAPTER SIX

Focusing Army Security Cooperation Activities to Cultivate Niche Capabilities

This chapter outlines a synchronized program of Army security cooperation activities to cultivate niche capabilities within non-core partner armies. Building on the previous chapter's discussion of ways to train, organize, and equip niche capability units in partner countries, this chapter provides a focused discussion of the means for doing so. It provides a rationale for focusing Army security cooperation activities to maximize impact, presents an illustrative approach for cultivating niche capabilities, and discusses the key challenges of building niche capabilities in non-core partner armies.

As discussed in previous chapters, the U.S. Army conducts a significant portion of DoD's security cooperation activities. However, in many cases Army security cooperation activities are currently not focused or sequenced in such a way as to maximize each event's contribution to Army Title 10 responsibilities. This chapter describes an approach to activity planning that can create synergies between DoD political-military requirements and the Army's Title 10 requirements. The Appendix discusses security cooperation terminology relative to Army security cooperation.

Focusing Army Security Cooperation Activities

As discussed in Chapter Five, Army planners currently lack a clear framework for focusing Army security cooperation activities on cultivating partner capabilities. As a result, activities are often conducted to serve the political-military requirements of DoD and the COCOMs, without considering HQDA's Title 10 objectives. Our research suggests that activity sequencing is important for five reasons. First, the progressive sequencing of Army security cooperation activities helps to lay the groundwork for more extensive capabilities-building training activities that may follow. Partner absorption rates are likely to increase if lower-impact activities are conducted first, followed by more challenging activities at a later point.

Second, conducting familiarizations and exchanges early on helps the U.S. Army to conduct "exploratory" events to determine the extent to which an investment in niche capabilities of the partner army is appropriate.

Third, activity sequencing allows HQDA to construct a more focused and tailored security cooperation package to build specific niche capabilities in a way that maximizes impact and leverages limited Army security cooperation resources. Such a package will assist HQDA in providing specific guidance to the ASCCs in the building of niche capabilities in non-core partner armies.

Fourth, activity sequencing provides HQDA with an approach to increase coordination of events and overall transparency with the ASCCs, COCOMs, other DoD security cooperation agencies, and within the U.S. Interagency to ensure that all relevant activities are leveraged for Title 10 requirements. Such an approach will help HQDA and the ASCCs to deconflict similar events and avoid repeating the same event year after year if effectiveness cannot be demonstrated.

Fifth, communicating the rationale and specific plans for activity sequencing to non-core partner armies provides them with visibility into U.S. expectations and plans for building niche capabilities in future years, reinforcing the niche concept in every activity conducted.

Figure 6.1 shows the strategic view for building niche capabilities in non-core partner armies. Non-capabilities-building programs (i.e., familiarizations and exchanges) are typically used in the beginning of a relationship with a partner country when bilateral ties are in an exploratory stage of development.

Figure 6.1
Sequencing of Security Cooperation: Strategic View

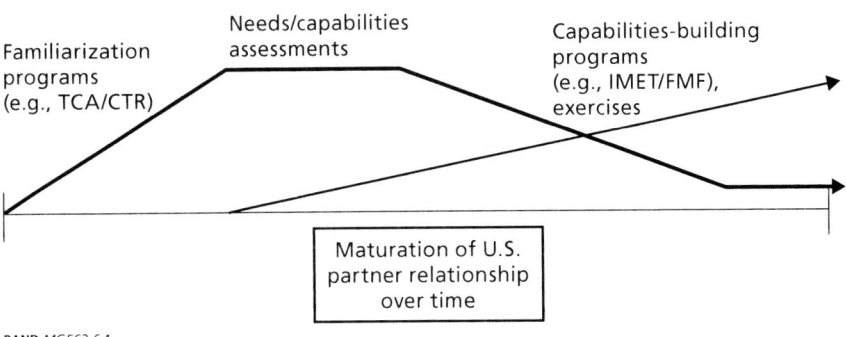

RAND *MG563-6.1*

Familiarization-type activities are typically phased out over time as the relationship matures. Next, prior to the provision of specific training and equipment, needs and capabilities assessments can be conducted to set the baseline requirements. Over time as the relationship matures, focused training and equipment can be provided to cultivate niche capabilities.[1]

The critical first step for HQDA is to provide strategic-level guidance that will direct the ASCCs and activity managers toward building compatibility with key non-core partner niche capabilities. As a rule, HQDA will likely want to provide broad strategic guidance (intent) to the ASCCs and activity managers in order to allow them the flexibility to amend and modify existing plans and activities to fulfill the Army and COCOM guidance. However, in some cases, HQDA may wish to specify particular activities and participants for Army security cooperation.

---

[1] U.S. law makes a distinction between international training programs and international familiarization programs. According to Army Regulation AR 350-17, there are three constituent elements to "training": (1) the action which the unit or soldier must be capable of performing; (2) the standard of performance (observable, measurable, and achievable) the unit or soldier must meet; and (3) the conditions under which the unit or soldier is expected to perform. In security cooperation, the determination of whether or not training of foreign forces has taken place is often difficult to make. Unless a proficiency assessment of some kind is given at the end of the event, training has not officially taken place.

At the present time, the five regional ASCCs (U.S. Army Europe—USAREUR; U.S. Army Central—USARCENT; U.S. Army South—USARSO; U.S. Army Pacific—USARPAC; and U.S. Army North—ARNORTH) primarily take their direction from the regional COCOMs, rather than from HQDA. They receive security cooperation resources from both the COCOMs and HQDA, with the exception of USARCENT, which is not a MACOM and receives resources from CENTCOM (with the exception of personnel and units). In practice, it is the COCOMs that provide the most specific security cooperation guidance by way of the Theater Security Cooperation Plans (TSCPs), Regional Strategies, and the Country Campaign Plans (CCPs). Because the ASCS uses the same overarching goals outlined in OSD's Security Cooperation Guidance, which is also what the COCOMs use in their TSCPs, at the top level the guidance is generally synchronized. It is in the country-specific guidance that the majority of differences occur, and HQDA does not provide guidance to ASCCs at the activity level. While the ASCCs all have their unique operating styles and peculiarities, as one might expect, all are required to execute guidance directed by the COCOMs and HQDA.

Against the backdrop of relations between HQDA and the ASCCs, how would progressive sequencing of activities work in practice? To what extent do planners consider the order of activities they are executing in relation to the goals they are trying to achieve? The study team found little evidence in the field or at HQDA that a conscious effort is made to sequence Army activities or DoD security cooperation activities on a broader scale. Perhaps this is the case on occasion, but this is not the standard—particularly with non-core partner armies that receive fewer security cooperation resources. There has not been a concerted effort to determine how the Army can use Army security cooperation activities to improve the effectiveness of capabilities of non-core partner armies and, as a result, there is no strategy in place for doing so at present.

Understanding the sequence by which activities should be conducted to cultivate coalition partners' niche capabilities may help to: (1) focus events and limited resources; (2) justify why events are conducted; (3) help deconflict between similar events; (4) produce a larger return on investment; and (5) demonstrate how an event is part of a larger niche capabilities-building plan. The next section describes the types of Army security cooperation activities that Army planners could use to cultivate niche capabilities in an effective and efficient manner.

Army Security Cooperation Activities Most Relevant to the Niche Capabilities MFC Strategy

Some activities reinforce certain HQDA security cooperation goals, while others are intended to provide security assistance in the form of training and equipment. Reinforcing activities, for example, include Army-to-Army staff talks, events conducted by DoD's regional centers.[2] Security assistance training and equipment can be provided through exercises, small unit exchanges, mobile training teams (MTTs), IMET, and FMF, for example. However, it is our assumption that the Army would benefit if all programs were to be incorporated into a broader planning framework that lays out the Army's specific equities and the goals it wants to achieve with each partner country.

In determining which activities would be the most effective in supporting the development and deployment of niche capabilities, the project team must consider questions regarding the appropriateness of available activities. For example, planners could consider questions such as: (1) To what extent can security cooperation activities be credited for building capabilities that were deployed by non-core partner armies to recent combat and stability and support operations? (2) Which activities were used to support coalition partners that deployed to Bosnia, Kosovo, Afghanistan, and Iraq? (3) Which ones appeared to have a greater impact, and why?

The study team identified several Army security cooperation activities that can provide the capabilities-building training and equipment necessary to build niche capabilities. Exercises, especially com-

---

[2] George C. Marshall Center for European Studies, Africa Center for Strategic Studies, Near East Center for Strategic Studies, the Asia-Pacific Center for Strategic Studies, and the Center for Hemispheric Defense Studies.

bined and Mission Rehearsal Exercises (MRXs), small unit exchanges, capabilities and needs assessments, MTTs, and the broader spectrum of security assistance (FMF, IMET, and EDA) are arguably the most important activities on the capabilities-building side of the spectrum. Each is discussed below with examples of specific, high-impact events.

Exercises and Small Unit Exchanges

There are many different kinds of bilateral and multilateral exercises. Mission Rehearsal Exercises (or pre-deployment exercises) and small unit exchanges (SUEs) are particularly important from the perspective of MFC planning. MRXs and SUEs represent the nexus between security cooperation and ongoing military operations. For example, MRXs conducted in Central Europe prior to a country's deployment to coalition operations in Iraq and Afghanistan not only provided for a "dress rehearsal" to test interoperability, but also allowed USAREUR and U.S. Air Force Europe (USAFE) a chance to conduct impromptu capabilities assessments of the individual units set to deploy. Capabilities-building bilateral and multilateral exercises are currently being conducted in every area of responsibility (AOR).[3]

Capabilities/Needs Assessments

Capabilities and needs assessments have been conducted by various DoD agencies (COCOMs, Defense Intelligence Agency, OSD, Joint Staff), but different approaches are utilized. For example, EUCOM conducts assessments of the defense and military establishments of its

---

[3]   USAREUR conducts both PfP and "In-the-spirit of" PfP bilateral multilateral exercises with the British, French, Germans, Poles, Romanians, and Bulgarians, just to mention a few. Improving compatibility is often the primary focus of such exercises. USARPAC's annual Multinational Planning Augmentation Team (MPAT) exercise is intended to improve the partners' collective ability to improve interoperability. PACOM's exercises help to address many of the interoperability challenges associated with deployment to a coalition operation for many of the Asia-Pacific partner armies. USARSO exercises have been used to prepare non-core partner armies for coalition operations. For example, the annual PKO NORTH computer-assisted exercise included critical training for coalition partners such as Honduras, El Salvador, Nicaragua, and the Dominican Republic, all of whom sent forces to Iraq. ARCENT conducts joint/coalition exercises, such as BRIGHT STAR, designed to increase regional involvement in pursuit of improved security and defense capabilities.

partner countries in Central Europe and Eurasia. These assessments have provided the United States with details of the level of competency of the force, equipment present, and training received. However, they have not properly assessed the strategic environment relative to the security needs of each partner.[4]

In contrast, SOUTHCOM has a more focused approach with respect to assessing the operational capabilities of coalition partners. SOUTHCOM J-5 (Strategy and Plans) responds to requests from partner countries in Latin America to conduct capabilities assessments of smaller pieces of the force structure. The purpose is to determine exactly what it would take to bring each capability to the point of being able to deploy it for a coalition operation. Approximately ten assessments have been conducted to date, including in Bolivia, Nicaragua, Colombia, and Ecuador. Capabilities in these countries have been assessed in terms of mobilization, training, infrastructure, intelligence, communications, sustainability, information operations, force protection, riverine, and civil affairs/civil-military operations.[5]

According to the SOUTHCOM J-5 staff, needs assessments are much more welcome in the Latin America AOR because they are focused on what the country requires to improve its own defense self-sufficiency, and such assessments are perceived as potentially bringing significant resources if a capability is deemed critical to coalition operations. By utilizing SOUTHCOM's approach to assessments, the Army is likely to gain information piecemeal as opposed to trying to conduct a full defense/military establishment assessment all at once. When the United States is perceived as a cooperative source of assistance, the partner's objections and defensive walls are lowered.[6]

---

[4]   Defense assessments to date have been conducted in all of the central and southeast European partners, as well as Georgia, Ukraine, Kazakhstan, Azerbaijan, and Moldova. The Georgia, Moldova, and Azerbaijan assessments were all-encompassing (i.e., assessment of the entire defense and military structure), while the Ukraine and Kazakhstan assessments focused on specific pieces of the force structure (for Ukraine, the Rapid Reaction Force was assessed, and for Kazakhstan, the Kazakh Peacekeeping Battalion was assessed).

[5]   Discussions with SOUTHCOM J-5 officials, Miami, FL, August 2003 and San Antonio, TX, January 2004.

[6]   Discussions at SOUTHCOM, January 2005.

Mobile Training Teams

In terms of mobile training teams (MTTs), a few good examples of how they were able to support the deployment of coalition partners include events with U.S. Marine Forces Europe that were conducted with Georgia to support the Georgia Train and Equip program (GTEP) and UH-1 helicopter programs. The MTT events helped prepare the Georgian forces for eventual deployment to Iraq.[7] Moreover, in Romania, a 2003 MTT event focused on the development of niche capabilities, specifically emphasizing mountain warfare special forces and military police, as well as medium airlift (C-130 aircraft). Like Georgia, Romania also deployed its special forces capabilities to Iraq. Interestingly, EUCOM and USAREUR have jointly supported the development of Romania's special forces (the Red Scorpions) through a coordinated approach focusing Joint Contact Team Program (JCTP) events, IMET training at U.S. facilities, and Joint Combined Exchange Training (JCETs) on this goal.[8]

Security Assistance (Title 22)

Security assistance continues to be the main vehicle by which non-core partner armies receive training and equipment. Most countries that are providing coalition support to OEF and OIF have received additional annual FMF for supporting the United States in the war on terrorism and to help build up their operational capabilities to these ends. Table 6.1 details some overall increases between 2001 and 2003 as illustrative examples.

Building Niche Capabilities Through Logically Sequenced Army Security Cooperation Activities

There is no shortage of Army security cooperation activities to draw upon in crafting a strategy to cultivate the capabilities of promising non-core partner armies. Given the requirements identified in Chapter

---

[7]   Although the primary purpose was to defeat the insurgency in the Pankisi Gorge region.

[8]   Discussions at EUCOM ECJ-5 and ECJ-4 staff, Stuttgart, Germany, May 2004.

Table 6.1
Selected FMF Allocations 2001–2003

Sample Country	Fiscal Year	FMF	Total Increase, 2001–2003
Uzbekistan	2001	1.7M	7.1M
	2002	25.2M	
	2003	8.8M	
Pakistan	2001	0	50M
	2002	0	
	2003	50M	
Georgia	2001	4.5M	2.5M
	2002	11M	
	2003	7M	
Philippines	2001	2M	18M
	2002	19M	
	2003	20M	
Jordan	2001	75M	123M
	2002	75M	
	2003	198M	

SOURCE: U.S. Department of State Security Assistance Database, which can be found on the Federation of American Scientists website, accessible online at http://www.fas.org/asmp/profiles/aid.

Five, Figure 6.2 lists both generic categories of activities (across the top) and specific examples of an activity that would be appropriate (across the bottom) to fulfill each requirement. The cultivation of constabulary forces is used as an illustrative example. These schemes are notional, but are intended to serve as a starting point for discussion among HQDA planners, program managers, and implementers on the importance of building upon past activities in working toward the goal of cultivating specific coalition partner capabilities.

Figure 6.2
Building Niche Capabilities in Constabulary Forces

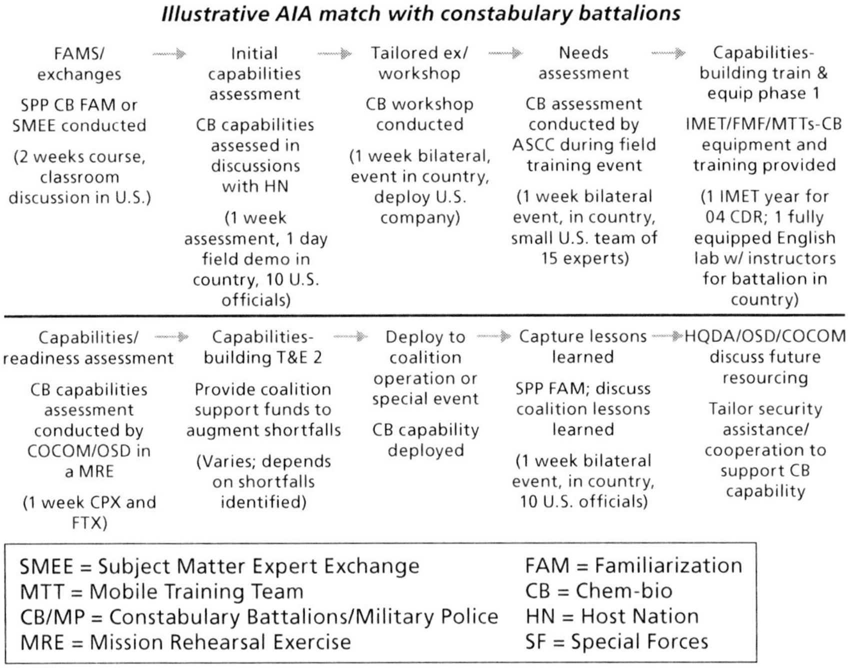

Illustrative AIA match with constabulary battalions

FAMS/ exchanges ⟶	Initial capabilities assessment ⟶	Tailored ex/ workshop ⟶	Needs assessment ⟶	Capabilities-building train & equip phase 1
SPP CB FAM or SMEE conducted (2 weeks course, classroom discussion in U.S.)	CB capabilities assessed in discussions with HN (1 week assessment, 1 day field demo in country, 10 U.S. officials)	CB workshop conducted (1 week bilateral, event in country, deploy U.S. company)	CB assessment conducted by ASCC during field training event (1 week bilateral event, in country, small U.S. team of 15 experts)	IMET/FMF/MTTs-CB equipment and training provided (1 IMET year for 04 CDR; 1 fully equipped English lab w/ instructors for battalion in country)

Capabilities/ readiness assessment ⟶	Capabilities-building T&E 2 ⟶	Deploy to coalition operation or special event ⟶	Capture lessons learned ⟶	HQDA/OSD/COCOM discuss future resourcing
CB capabilities assessment conducted by COCOM/OSD in a MRE (1 week CPX and FTX)	Provide coalition support funds to augment shortfalls (Varies; depends on shortfalls identified)	CB capability deployed	SPP FAM; discuss coalition lessons learned (1 week bilateral event, in country, 10 U.S. officials)	Tailor security assistance/ cooperation to support CB capability

SMEE = Subject Matter Expert Exchange	FAM = Familiarization
MTT = Mobile Training Team	CB = Chem-bio
CB/MP = Constabulary Battalions/Military Police	HN = Host Nation
MRE = Mission Rehearsal Exercise	SF = Special Forces

RAND *MG563-6.2*

Complementary Approaches to Developing Niche Capabilities

There are complementary approaches that can usefully support the Niche Capability Planning Framework. First, as discussed in Chapter Three, the Army could promote the development of *dual-use niche capabilities*, i.e., capabilities that are useful both domestically and for coalition operations, to address national and regional security needs. Second, the Army could also promote the development of *combined dual-use niche capabilities*, which focus on the forces of two or more countries to address regional or transnational security problems. Third, the Army could work through the COCOMs to leverage the experience of the more well-developed and promising allies as "enabling partners" to share the burden and free up U.S. forces. Each option is discussed below.

First, for the non-core partner armies, especially those facing a domestic security threat, it would be useful if the niche capabilities developed have a dual-use, that is, they are designed to be useful at home as well as abroad. Examples of such capabilities include: military police/constabulary forces, chemical-biological defense and consequence management, engineering, emergency medical, civil affairs, explosive ordnance disposal, humanitarian demining, and peacekeeping. There is also a greater chance the partner will be committed to developing and sustaining its niche capabilities if it can also be used for domestic purposes or as a potential revenue-generator, such as peacekeeping units for UN operations.[9]

Second, another option for the United States to consider is the cultivating of multilateral or *combined dual-use niche capabilities* among several non-core partner armies within a given region. Some examples of possible combined dual-use niche capabilities include military police, peacekeeping/peace support operations, explosive ordnance disposal, and demining teams. Such capabilities could serve as a catalyst for greater regional cooperation, leverage U.S. security cooperation resources, and possibly attract support from other donor countries and multilateral organizations. In the case of Eastern Europe and Eurasia, such an approach could lead to additional resources from NATO, the European Union (EU), and the Organization for Security and Cooperation in Europe (OSCE). For example, in Central Asia, Kazakhstan, Kyrgyzstan, and Tajikistan have indicated a desire to develop deployable peacekeeping capabilities as well as a regional disaster response capability.[10] Several Central American partners have a similar desire to

---

[9] Several Latin American and Asia-Pacific partners, for example, are interested in peacekeeping in part for the monetary payoff that ensues when deployed for UN missions.

[10] Encouraging the development of such capabilities between these countries could serve many purposes, not least promoting regional cooperation between two countries that are typically distrustful of one another, such as in Central Asia and Latin America.

establish a deployable regional disaster response capability.[11] In Latin America, there are also opportunities for developing combined dual-use niche capabilities, especially in the areas of military police, consequence management/disaster relief, and combat search and rescue.[12] In the Asia-Pacific region, consequence management, engineering, emergency medical, and peacekeeping are all viable options.[13] Combined capabilities such as these and also other dual-use capabilities have the potential to encourage regional cooperation and security at home, and could also increase the effectiveness and efficiency of the coalition.

Third, another model to consider is to make better use of *enabling partner countries* to share their experience, expertise, and facilitate favorable relations in the region. These might include civil-military cooperation (CIMIC) capabilities of the Netherlands or Denmark, special forces capabilities of Poland or Romania, or as mentioned above, the carabinieri skills of the Italians[14] and gendarme capabilities of the French. The newest NATO members in particular have considerable experience and lessons learned in defense and military reform efforts that they could share with Partnership for Peace members to the east in the South Caucasus and Central Asia.[15]

Other potential models for utilizing niche capabilities could include their employment for special international events, such as the Olympic Games, or regional special events. A recent example of this would be the request of the government of Greece in March 2004 to the Czech Republic for use of its capable chemical and biological

---

[11] Per discussions with SOUTHCOM, USARSO, and G-35 officials.

[12] Per discussions with SOUTHCOM, USARSO, and G-35 officials.

[13] Per discussions with PACOM and USARPAC official, Honolulu, HI, September 2003.

[14] Through the U.S. Global Peacekeeping Initiative Funds, the United States is currently funding the Italians to train carabinieri skills to foreign forces at the school in Vincenza, Italy.

[15] For example, the Baltic countries (Estonia, Latvia, and Lithuania) have instituted a lessons learned sharing forum with the three South Caucasus countries (Georgia, Azerbaijan, and Armenia).

defense battalion during the Olympic Games in Athens.[16] Not only did this free up U.S. and other NATO allied forces, but also helped the Czech Republic build its international prestige and competency in this particular niche area. This is also the ideal model for burden sharing among NATO allies within the context of the NATO Response Force (NRF).

Key Challenges to Building and Sustaining Niche Capabilities

There are several challenges associated with focusing Army security cooperation activities on building niche capabilities in coalition partners that the Army will need to address as it develops the niche capabilities strategy. This section discusses the largest and most important of these challenges.

HQDA Strategic Planning and Links to the COCOM/ASCC

Given the cross-cutting nature of security cooperation programs executed by HQDA, OSD/Joint Staff, COCOMs, ASCCs, and other DoD agencies with non-core partner armies, increasing visibility into the strategic planning process of all the players is critical in terms of understanding where HQDA can effectively leverage other, related programs for Title 10 requirements. One possibility would be for Army planners to regularly take part in COCOM Theater Security Cooperation Working Groups, where the majority of strategic planning for security cooperation takes place. Another possibility would be for HQDA to provide tailored guidance to the ASCCs to build niche capabilities. It is important for HQDA to understand the nuances among the COCOMs and ASCCs in the security cooperation planning process, which is explained below.

Strategic planning across the COCOMs is now fairly standardized between the Theater Security Cooperation Plans, regional strat-

---

[16] Thomas Ricks, "NATO Pledges Ships and Aircraft to Help Safeguard Olympics," *The Washington Post*, 27 March 2004, p. A14.

egies, and the Country Campaign Plans, but relations between the COCOMs and the ASCCs vary greatly. None of the TSC strategies emphasize the building of niche capabilities as a goal, but all make reference to the need to cultivate the capabilities of coalition partners, per guidance in the QDR, SCG, and ASCS.

EUCOM's TSC emphasis is based in part on enabling partners, as discussed above, as well as enabling programs,[17] where DoD encourages key allies to take on extra responsibilities with regard to promising but less capable coalition partners in the Middle East, Caucasus, and Africa.[18]

In PACOM's TSCP, although the development of effective multilateral coalitions is a distinct goal, the strategy does not specify how this goal will be accomplished and through which activities.[19] SOUTHCOM's TSCP focuses on the capabilities-building aspect of security cooperation. Its Prioritized Capabilities and Tasks Lists (PCTLs) discuss the need to build capabilities that are interoperable with the United States and other coalition partners. SOUTHCOM focuses on building capabilities for peacekeeping and peace support operations, as well as on developing effective police forces. Defense self-sufficiency is key in this AOR. SOUTHCOM also links security cooperation activities to large ungoverned territories, such as the Andean Ridge.[20]

EUCOM and CENTCOM are attempting to prioritize resources based on the ability of the partner to contribute meaningful, needed

---

[17] There is a COCOM-wide recognition that DoD cannot do everything and therefore, where appropriate, should coordinate U.S. security cooperation goals and activities with key NATO allies to encourage donor countries to step in where they have the appropriate desire and skills.

[18] Insights acquired during one author's participation in EUCOM's Theater Security Cooperation workshop, Heidelberg, Germany, March 2004.

[19] PACOM is an interesting case in the area of developing measures of effectiveness that link to its Theater Security Cooperation Management Information System (TSCMIS). PACOM J-5 has established an online survey that program and activity managers are required to fill out. This survey focuses on two questions: (1) are we doing the right things? and (2) are we doing things right? PACOM involves the component commands, specifically USARPAC, which is responsible for providing information to populate the TSCMIS system.

[20] Discussions with SOUTHCOM officials, Miami, FL, August 2004.

capabilities to NATO and coalition operations, which is evident in each of their TSCPs. However, SOUTHCOM and PACOM have not systematically incorporated this goal into their TSCP planning processes.

HQDA would benefit in terms of its strategic planning by accessing TSC plans and databases to learn what other COCOM activities contribute to building niche capabilities. Moreover, a more active HQDA participation in COCOM and ASCC planning processes, such as the COCOM TSC Working Groups and Security Assistance Conferences, would allow for increased Army input into COCOM plans in terms of building niche capabilities in each respective AOR. Overall, a greater understanding of the nuances of each COCOM and ASCC (e.g., security cooperation planning, execution, capabilities-building priorities, and after action reporting) could increase HQDA's ability to leverage ongoing activities to support Army Title 10 requirements.

Conclusions

Theater security cooperation is now being viewed as a way to develop allied and friendly military capabilities, or build partner capacity, for coalition operations; however, the explicit connection between security cooperation activities and coalition support is still muddled in the guidance documents and in security cooperation planning and execution. Ultimately, Army planners require a framework for working more effectively with specific non-core partners that simultaneously addresses Army Title 10 requirements and broader U.S. national security objectives. In the Appendix, additional sequencing schemes are presented to help Army planners focus activities on developing specific niche capabilities. The schemes will ensure that each activity conducted by the Army is used to its maximum benefit and will help the Army focus its approach to enhance the effectiveness of future coalition operations with non-core partners.

CHAPTER SEVEN

Conclusions and Recommendations

The contemporary security environment poses major challenges for the Army as it works to build MFC with partner armies. During the Cold War, key partner armies were linked tightly to the U.S. Army by longstanding alliance relationships. Army planners had a clear idea of where potential operations would likely occur, which multinational partners would participate, and what type of military operations they would conduct together. In the post-9/11 era, and particularly in the context of the turn toward coalitions of the willing, Army planners face a much less predictable planning environment. They cannot confidently forecast the location of future contingencies, the nature of future operations, or which armies, beyond a handful of extraordinary allies, might participate in a future coalition operation. The challenges facing Army planners have therefore become much more complex and difficult in recent years.

To better manage this increasing complexity, the Army created an ASCS to focus and rationalize the scarce resources available for building compatibility with partner armies. Its purpose is to generate and promulgate Army security cooperation priorities throughout the service and within the broader community of the Department of Defense and U.S. government agencies conducting security cooperation with foreign governments.

The purpose of this study is to help the Army create a conceptual framework for MFC planning in the contemporary security environment, particularly with regard to non-core partner armies. We recommend that the Army consider adopting a framework for building niche capabilities with non-core partner armies. This framework will incorporate, at a minimum, the Title 10 needs of the Army, the political characteristics of potential partners, and the resources required to cultivate niche capabilities in selected partners. This study proposes such a framework, in the form of the Niche Capability Planning Framework. It also makes a number of specific recommendations to Army planners regarding the security cooperation planning, as well as policy and process challenges. These two sets of recommendations are articulated below.

Recommendations

On the basis of our analysis of the Army's current security cooperation planning process, we offer the following six recommendations.

Incorporate a Niche-Based Approach into MFC Planning and the ASCS

The Army took a major step in developing a security cooperation plan. Our assessment suggests that a niche-based approach may be appropriate to security cooperation planning with non-core partner armies. We recommend that the Army consider incorporating the niche approach into MFC planning and the ASCS. This guidance should be addressed to the Army service component commands and Army major commands. It should specify the niche capabilities the Army seeks to develop, the partners the Army wishes to work with in this endeavor, and the security cooperation resources that will be devoted to the project.

Adopt a Niche Capability Planning Framework

To develop a niche-based framework with non-core partner armies, the Army should embrace a deliberative planning framework that considers, at a minimum, the Title 10 needs of the Army, the political characteristics of potential partners, and the resources required to cultivate niche capabilities in selected partners. We have outlined such a construct, which we term the Niche Capability Planning Framework.

Consider MFC as a Goal for Army Security Cooperation Activities

Currently, the Army thinks about multinational force compatibility as a category of activities. It may be more appropriate to think of MFC as a goal rather than a set of activities. Throughout this report, we have referred to "MFC-oriented activities" or to the goal of "building MFC" rather than to "MFC activities." This usage is more accurate and recognizes the fact that any activity can make a contribution to MFC.

At the same time, our analysis suggests that the definition of "multinational force compatibility" in the current AR 34-1 is probably too broad to serve Army planners in their duties. AR 34-1 currently defines MFC as "the collection of capabilities, relationships, and processes that together enable the Army to conduct effective multinational operations across the full spectrum of military missions. It encompasses not only the capability to conduct effective military operations with coalition partners, but also the factors that contribute to the development and maintenance of an alliance or coalition relationship."[1] A revised definition, focusing on the outcome of MFC (i.e., MFC as a goal) would help reduce confusion and would fit better as a subset of the ASCS. We suggest that the definition simply promote MFC as one of the key goals of Army security cooperation activities.

Increase HQDA Awareness of Activities Conducted in Theater

Through the implementation of the Army International Affairs Knowledge Sharing System (AIAKSS) and ability to leverage the respective COCOM TSCMIS databases, HQDA is steadily developing a better handle on the specific activities executed under Army security cooperation programs. This is an important starting point, but not the end-state. The development of multiple databases of activities and broader DoD security cooperation activities creates opportunities for HQDA to increase its visibility into the specific activities conducted globally. It is important for HQDA to continue to track the COCOM/ASCC security cooperation databases in an effort to increase transparency into those activities conducted that build partner country capabilities.

---

[1] AR 34-1, para. 2-1.

Leverage DoD, Interagency, and Allied Activities That Build Niche Capabilities

Army planners should also seek to leverage other programs and activities within the USG to the benefit of its Title 10 requirements. The Army Staff will logically focus efforts on activities the Army can directly influence (resource/policy oversight), but other DoD activities should also be monitored (including those of other services, OSD, the Joint Staff, DoD agencies such as the Defense Threat Reduction Agency, or DTRA). Eventually, the Army will want to extend its vision to encompass broader USG Interagency programs. Finally, the activities conducted by some U.S. allies, particularly related to the development of niche capabilities, should also be taken into account, where and when possible.

Recognize Non-Core Partners as Opportunities

Finally, on a broader note, Army planners should recognize that they have the ability to determine whether non-core partner armies will be problems or assets for future warfighting commanders. When non-core partner contingents show up for coalition operations, they often create significant challenges for U.S. forces because their doctrine, organization, and equipment are not compatible with U.S. capabilities. This may always be the case for some non-core partners. However, if Army planners can implement a proactive niche capabilities strategy via the ASCS, they can help ensure that many non-core partners that participate in future operations will bring capabilities that are useful to U.S. warfighters. In so doing, the Army might move some distance toward relieving some of its HD/LD burdens in a cost-effective manner and toward helping partner countries develop capabilities that will be useful in times of domestic stress. It might well be argued that not only is the niche-based strategy a smart investment for the Army, but in the current security environment it is also a Title 10 necessity.

Defining Security Cooperation Terminology

What differentiates security cooperation programs from funding sources, initiatives, activities/events, and relationships? These are important distinctions. The misuse of these terms causes confusion in strategy formulation and policy coordination. This appendix defines and contrasts these often-misunderstood concepts in order to provide a clear conceptual basis for the Niche Capability Planning Framework.

Funding Sources

Funding sources are large umbrella resource streams that fund initiatives or programs. The Freedom Support Act (FSA), which funds many initiatives and programs in Eurasia, is an example of a funding source. FSA provides funding, for example, to the State Department's Export Control and Related Border Security (EXBS) program.

Initiatives/MDEPs

Initiatives or Management Decision Packages (MDEPs) are funding sources for a collection of programs that pursue a particular set of goals. Examples of initiatives include the Warsaw Initiative Fund (WIF), which funds programs in central and southern Europe as well as Eurasia, and the Cooperative Threat Reduction (CTR) Defense and Military Contacts program. CTR and WIF fund some Army security cooperation activities, including the National Guard's State Partnership Program (SPP).

Programs

Programs are a set of activities or events coordinated to achieve a certain set of objectives. Programs have the following defining characteristics at a minimum:

- Mission and set of specific objectives;
- Activities or events executed;
- Manager(s) for policy and/or resource oversight; and
- Reporting requirements to an oversight agency or office.

There are two kinds of DoD and Army security cooperation programs: *independent* and *dependent* programs. Independent programs have their own line item in a budget (e.g., the Army POM) and therefore are fiscally secure and do not have to solicit funds from other sources to execute activities. An example of an independent program is EUCOM's Joint Contact Team Program (JCTP). Another is DTRA's International Counterproliferation Program (ICP).

Dependent programs rely on initiatives or other programs for funding. Nearly all Army security cooperation activities are dependent programs. An example is the Partnership for Peace Information Management System (PIMS), which is funded and overseen by the State Department under the Warsaw Initiative Fund, but executed by DoD. Another is the Civil Military Emergency Preparedness (CMEP) program, funded by a variety of sources. Since Army forces are involved in the execution of PIMS, CMEP, and other dependent program activities, HQDA claims these as its own. Other examples include "in the spirit of" Partnership for Peace exercises, the Army's chaplain programs, and Army medical events.

It is also noteworthy that there may be different offices or individuals responsible for policy and planning, resource management, and program execution within organizations and at different organizational levels. Examples are the FMF, FMS, and IMET, all of which are executed by DoD but funded and overseen by the State Department.

Activities/Events

Activities and events are actions (directed, funded, and/or supervised by program managers) that programs support through implementation, support, and/or funding. Activities are generic (e.g., Army-to-Army staff talks), while events are specific (e.g., U.S.-Czech Republic Army-to-Army staff talks).

Table A.1 summarizes the distinctions of the concepts discussed above.

Table A.1
Building the Distinctions

Term	Defining Characteristics	Example
Funding source	Money	Freedom Support Act
Initiative	Money and broad goals	Warsaw Initiative
Program	Specific mission/objectives, manager, activities, reporting requirements	Civil Military Emergency Preparedness
Activities/events	Resourced, actionable, designed to address specific objectives and should be executed as part of a larger strategy	Joint Staff regional disaster response exercise
Relationships	A collection of organizations that coordinate their programs and often draw from the same funding sources to support their activities	National Guard's State Partnership Program

Relationships

Although more difficult to define than other categories, relationships can be viewed as a collection of organizations that coordinate their programs and often draw from the same funding sources to support their activities, but also draw from a variety of other sources in the public and private sectors. An example is the National Guard Bureau's State Partnership Program, considered an Army security cooperation activity because the active Army executes SPP events. SPP relies on different funding sources and independent programs for its resources, but it also has a relatively small amount of dedicated resources (approximately $1.5 million annually) called Minuteman Fellowships. SPP is allowed to accept grants to execute its military-military, civil-military, and civil-civil activities. SPP creates relationships with specific partners that involve all these resources.

In practice, funding sources or initiatives are often labeled as programs. For example, the Department of State's Anti-Terrorism Assistance (ATA) program is better described as an initiative, as it is a funding source with a specific mission, which is to build up the counterterrorism capabilities of key allies and partners.

Adding to the confusion, sometimes programs are also funding sources. An example would be the DoD Combating Terrorism Fellowships Program (CTFP), which has the mission to improve the combating terrorism skills of individuals from key coalition partners, activities such as training courses, and a managing office in OSD. But it is also viewed as a funding source because it supports other programs and activities, such as the National Guard Border Security program in Ukraine. Figure A.1 is an example of the logical flow from funding source to initiative to program to activity/event, including executive oversight.

Figure A.1
Example of the Logical Flow from Funding Source to Initiative to Program to Activity/Event

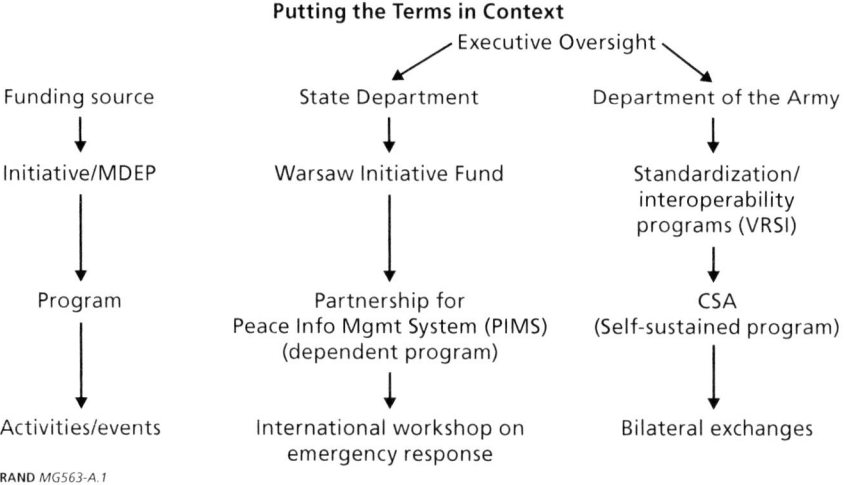

RAND *MG563-A.1*

Bibliography

Books, Reports, Articles, Journals

Chu, David, "Prepared Statement of the Honorable David S. Chu, Under Secretary of Defense (Personnel and Readiness) Before the House Armed Services Committee," 5 November 2003.

Davis, Lynn, et al., *Stretched Thin: Army Forces for Sustained Operations*, Santa Monica, CA: RAND Corporation, 2005. As of 9 January 2007: http://www.rand.org/pubs/monographs/MG362/

Dempsey, Judy, "Poland Is Squeezed by Cost of Iraq Role," *International Herald Tribune*, 18 October 2004.

Farrell, Theo, and Terry Terriff (eds.), *The Sources of Military Change: Culture, Politics, Technology*, Boulder, CO: Lynne Rienner, 2002.

"Feith: No U.S. Bases to Be Built in Eastern Europe," Defense Alert, *InsideDefense* (online at InsideDefense.com), 23 June 2004.

Grissom, Adam, Nora Bensahel, John Gordon, Terrence K. Kelly, and Michael Spirtas, *U.S. Army Transformation and the Future of Coalition Warfighting*, Santa Monica, CA: RAND Corporation, 2006.

Jablonsky, David, "Army National Guard Leaves USAF Gates," *Air Force Print News*, April 2003.

Jones, A. Elizabeth, Assistant Secretary for European and Eurasian Affairs, Department of State, "The Road to NATO's Prague Summit: New Capabilities, New Members, New Relationships," speech, 17 January 2002. As of 9 January 2007: http://state.gov/p/eur/rls/rm/2002/

Jones, Archer, *The Art of War in the Western World*, Champaign-Urbana: University of Illinois Press, 2001.

Milbank, Dana, "Uzbekistan Thanked for Role in War: U.S., Tashkent Sign Cooperation Pact," *The Washington Post*, 15 March 2002.

Moroney, Jennifer, "Western Security Cooperation with Central Asia: CIS or the European Security Order?" in Graeme Herd and Jennifer Moroney, *Security Dynamics in the Former Soviet Bloc*, London: Routledge/Curzon, 2003.

Nardulli, Bruce, "The U.S. Army and the Offensive War on Terrorism," in Lynn Davis and Jeremy Shapiro (eds.), *The U.S. Army and the New National Security Strategy*, Santa Monica, CA: RAND Corporation, 2003, pp. 27–58. As of 9 January 2007:
http://www.rand.org/pubs/monograph_reports/MR1657/

O'Hanlon, Michael, "Rebuilding Iraq and Rebuilding the U.S. Army," Saban Center Middle East Memo Number 3, 4 June 2004. As of 27 October 2004:
http://www.brookings.edu/views/op-ed/ohanlon/20040604.htm

Oliker, Olga, et al., *Aid During Conflict: Interaction Between Military and Civilian Assistance Providers in Afghanistan, September 2001–June 2002*, Santa Monica, CA: RAND Corporation, 2004. As of 9 January 2007:
http://www.rand.org/pubs/monographs/MG212/

Palmer, Adele R., and David J. Osbaldeston, *Incremental Costs of Military and Civilian Manpower in the Military Services*, Santa Monica, CA: RAND Corporation, 1988. As of 9 January 2007:
http://www.rand.org/pubs/notes/N2677/

Ricks, Thomas, "NATO Pledges Ships and Aircraft to Help Safeguard Olympics," *The Washington Post*, 27 March 2004, p. A14.

Russett, Bruce, "The Fact of the Democratic Peace, and Why Democratic Peace," in Michael E. Brown, Sean M. Lynn-Jones, and Steven E. Miller (eds.), *Debating the Democratic Peace*, Cambridge, MA: MIT Press, 1996.

Simon, Michael, and Erik Gartzke, "Political System Similarity and the Choice of Allies: Do Democracies Flock Together or Do Opposites Attract?" *Journal of Conflict Resolution*, Vol. 40, No. 4, December 1996, pp. 617–635.

Sortor, Ronald E., and J. Michael Polich, *Deployments and Army Personnel Tempo*, Santa Monica, CA: RAND Corporation, 2002. As of 9 January 2007:
http://www.rand.org/pubs/monograph_reports/MR1417/

Szayna, Thomas, Frances Lussier, Krista Magras, Olga Oliker, Michele Zanini, and Robert Howe, *Improving Army Planning for Future Multinational Coalition Operations*, Santa Monica, CA: RAND Corporation, 2001. As of 9 January 2007:
http://www.rand.org/pubs/monograph_reports/MR1291/

Walt, Stephen, Kenneth Waltz, Carol Ember, Melvin Ember, and Bruce Russett, "Peace Between Participatory Polities," *World Politics*, July 1992.

U.S. Department of Defense, *Annual Defense Review*, Washington, D.C.: Government Printing Office, 2004.

Focused Discussions

Chem-bio defense specialist, U.S. Department of the Army G-35 DAMO-SSI, discussion with the authors, June 2004.

Discussions in the Joint Staff, J-5, January 2005.

Discussions with selected defense officials and military scholars, Kyiv, Ukraine, February 2005.

Discussions with European and Eurasia defense officials, March 2005.

Discussion with Joint Staff official, March 2005.